Introduction to Solar Photovoltaics

Trainee Guide
First Edition

Prentice Hall

Boston   Columbus   Indianapolis   New York   San Francisco   Upper Saddle River
Amsterdam   Cape Town   Dubai   London   Madrid   Milan   Munich   Paris   Montreal   Toronto
Delhi   Mexico City   Sao Paulo   Sydney   Hong Kong   Seoul   Singapore   Taipei   Tokyo

National Center for Construction Education and Research

President: Don Whyte
Director of Product Development: Daniele Stacey
Introduction to Solar Photovoltaics Project Manager: Matt Tischler
Production Manager: Tim Davis
Quality Assurance Coordinator: Debie Ness
Editor: Rob Richardson
Desktop Publishing Coordinator: James McKay
Production Assistant: Laura Wright

NCCER would like to acknowledge the contract service provider for this curriculum:
Topaz Publications, Syracuse, New York.

This information is general in nature and intended for training purposes only. Actual performance of activities described in this manual requires compliance with all applicable operating, service, maintenance, and safety procedures under the direction of qualified personnel. References in this manual to patented or proprietary devices do not constitute a recommendation of their use.

10 9 8 7 6

Prentice Hall
is an imprint of

www.pearsonhighered.com

ISBN 13: 978-0-13-213726-3

Preface

To the Trainee

The solar power industry is a rapidly growing field that is expected to help ease human dependence on the use of fossil fuels. Solar panels are now rated to produce up to 600 volts of electricity, and the cost of purchasing and installing these panels for residential purposes has been reduced considerably. Because of this, the need for solar photovoltaic installers has increased and is projected to grow with the demand for solar installations.

Through both government and private initiatives, the increased need for qualified solar photovoltaic installers will provide you with an opportunity to join an expanding and viable workforce. According to the Energy Information Administration, domestic shipments of solar panels have increased by nearly 2400 percent since 2000, and the U.S. Department of Labor expects the solar photovoltaic workforce to grow by 50 percent in the coming years.

This module covers a broad range of information that is valuable to entry-level photovoltaic installers, including system components, design considerations, environmental effects, and site analysis. *Introduction to Solar Photovoltaics* is intended to provide trainees with the requisite knowledge to pass the North American Board of Certified Energy Practitioners (NABCEP) PV Entry Level Exam.

By taking this course, you're preparing yourself with the knowledge necessary to be part of the solar energy workforce. Today, the growth of power generation is tied to clean and renewable energy. The skills you learn in this module will help provide you with a lifetime of opportunity in this field.

We invite you to visit the NCCER website at www.nccer.org for the latest releases, training information, newsletter, and much more. You can also reference the Pearson product catalog online at www.nccer.org.

Your feedback is welcome. You may email your comments to curriculum@nccer.org or send general comments and inquiries to info@nccer.org.

NCCER Standardized Curricula

NCCER is a not-for-profit 501(c)(3) education foundation established in 1995 by the world's largest and most progressive construction companies and national construction associations. It was founded to address the severe workforce shortage facing the industry and to develop a standardized training process and curricula. Today, NCCER is supported by hundreds of leading construction and maintenance companies, manufacturers, and national associations. The NCCER Standardized Curricula was developed by NCCER in partnership with Pearson Education, Inc., the world's largest educational publisher.

Some features of NCCER's Standardized Curricula are as follows:

- An industry-proven record of success
- Curricula developed by the industry for the industry
- National standardization providing portability of learned job skills and educational credits
- Compliance with Office of Apprenticeship requirements for related classroom training (CFR 29:29)
- Well-illustrated, up-to-date, and practical information

NCCER also maintains a National Registry that provides transcripts, certificates, and wallet cards to individuals who have successfully completed modules of NCCER's Standardized Curricula. *Training programs must be delivered by an NCCER Accredited Training Sponsor in order to receive these credentials.*

Contren® Curricula

NCCER's training programs comprise nearly 80 construction, maintenance, pipeline, and utility areas and include skills assessments, safety training, and management education.

Boilermaking
Cabinetmaking
Carpentry
Concrete Finishing
Construction Craft Laborer
Construction Technology
Core Curriculum:
  Introductory Craft Skills
Drywall
Electrical
Electronic Systems Technician
Heating, Ventilating, and
  Air Conditioning
Heavy Equipment Operations
Highway/Heavy Construction
Hydroblasting
Industrial Coating and Lining
  Application Specialist
Industrial Maintenance
  Electrical and
  Instrumentation Technician
Industrial Maintenance
  Mechanic
Instrumentation
Insulating
Ironworking
Masonry
Millwright
Mobile Crane Operations
Painting
Painting, Industrial
Pipefitting
Pipelayer
Plumbing
Reinforcing Ironwork
Rigging
Scaffolding
Sheet Metal
Site Layout
Sprinkler Fitting
Tower Crane Operator
Welding

Green/Sustainable Construction

Your Role in the Green
  Environment
Sustainable Construction
  Supervisor
Introduction to Weatherization
Weatherization Installer

Energy

Introduction to the Power
  Industry
Power Industry Fundamentals
Power Generation Maintenance
  Electrician
Power Generation I&C
  Maintenance Technician
Power Generation Maintenance
  Mechanic
Steam and Gas Turbine
  Technician
Introduction to Solar
  Photovoltaics
Introduction to Wind Energy

Pipeline

Control Center Operations,
  Liquid
Corrosion Control
Electrical and Instrumentation
Field Operations, Liquid
Field Operations, Gas
Maintenance
Mechanical

Safety

Field Safety
Safety Orientation
Safety Technology

Management

Introductory Skills for the
  Crew Leader
Project Management
Project Supervision

Supplemental Titles

Applied Construction Math
Careers in Construction
Tools for Success

Spanish Translations

Basic Rigging
  (Principios Básicos de
  Maniobras)
Carpentry Fundamentals
  (Introducción a la
  Carpintería, Nivel Uno)
Carpentry Forms
  (Formas para Carpintería,
  Nivel Trés)
Concete Finishing, Level One
  (Acabado de Concreto,
  Nivel Uno)
Core Curriculum:
  Introductory Craft Skills
  (Currículo Básico:
  Habilidades Introductorias
  del Oficio)
Drywall, Level One
  (Paneles de Yeso, Nivel Uno)
Electrical, Level One
  (Electricidad, Nivel Uno)
Field Safety
  (Seguridad de Campo)
Insulating, Level One
  (Aislamiento, Nivel Uno)
Masonry, Level One
  (Albañilería, Nivel Uno)
Pipefitting, Level One
  (Instalación de Tubería
  Industrial, Nivel Uno)
Reinforcing Ironwork, Level One
  (Herreria de Refuerzo,
  Nivel Uno)
Safety Orientation
  (Orientación de Seguridad)
Scaffolding
  (Andamios)
Sprinkler Fitting, Level One
  (Instalación de Rociadores,
  Nivel Uno)

Acknowledgments

This curriculum was revised as a result of the farsightedness and leadership
of the following sponsors:

Florida Solar Energy Center
Industrial Management & Training Institute,
Inc. of Connecticut
Marion Technical Institute
NeoSolvis Engineering C.S.P

Pumba Electric LLC
Solar Source Institute
Tri-City Electrical Contractors, Inc.
Westside Technical Center

This curriculum would not exist were it not for the dedication and unselfish energy of those volunteers
who served on the Authoring Team. A sincere thanks is extended to the following:

Nicolás Estévez
L. J. LeBlanc
Joseph S. Lowe
Mark Maher

Mike Powers
Susan Schleith
Antonio Vazquez
Marcel Veronneau

NCCER Partners

American Fire Sprinkler Association
Associated Builders and Contractors, Inc.
Associated General Contractors of America
Association for Career and Technical Education
Association for Skilled and Technical Sciences
Carolinas AGC, Inc.
Carolinas Electrical Contractors Association
Center for the Improvement of Construction
   Management and Processes
Construction Industry Institute
Construction Users Roundtable
Design Build Institute of America
Merit Contractors Association of Canada
Metal Building Manufacturers Association
NACE International
National Association of Manufacturers
National Association of Minority Contractors
National Association of Women in Construction
National Insulation Association
National Ready Mixed Concrete Association
National Technical Honor Society
National Utility Contractors Association

NAWIC Education Foundation
North American Technician Excellence
Painting & Decorating Contractors of America
Portland Cement Association
SkillsUSA
Steel Erectors Association of America
U.S. Army Corps of Engineers
Women Construction Owners & Executives, USA
University of Florida, M.E. Rinker School of
   Building Construction

Contents

Introduction to Solar Photovoltaics

57101-11

Trainees with successful module completions may be eligible for credentialing through NCCER's National Registry. To learn more, go to **www.nccer.org** or contact us at **1.888.622.3720.** Our website has information on the latest product releases and training, as well as online versions of our Cornerstone newsletter and Pearson's Contren® product catalog.

Your feedback is welcome. You may email your comments to **curriculum@nccer.org,** send general comments and inquiries to **info@nccer.org**, or use the User Update form at the back of this module.

 V.1 4/11

Objectives

When you have completed this module, you will be able to do the following:

1. Identify photovoltaic (PV) applications and advantages.
2. Identify system components and their functions.
3. Identify safety hazards associated with PV installations.
4. Trace a basic electrical circuit and perform calculations using Ohm's law.
5. List PV system sizing considerations.
6. Identify PV electrical and mechanical system design considerations.
7. Describe the tasks required to complete a site analysis.
8. Identify the effects of the environment on panel output.
9. Describe how to install a simple grid-connected PV system.
10. Explain how to assess system operation and efficiency.
11. Recognize the tasks required when performing PV maintenance and troubleshooting.
12. Identify appropriate codes and standards concerning installation, operation, and maintenance of PV systems and equipment.

Performance Tasks

This is a knowledge-based module; there are no performance tasks.

Trade Terms

Air mass
Altitude
Ambient temperature
Amorphous
Array
Autonomy
Azimuth
Backfeed
Balance of system (BOS)
Brownout
Building-integrated photovoltaics (BIPV)
Bypass diode
Charge controller
Combiner box
Concentrator
Declination
Depth of discharge (DOD)

Doped
Dual-axis tracking
Electrochemical solar cells
Elevation
Fuel cells
Grid-connected system
Grid-interactive system
Grid-tied system
Heat fade
Hybrid system
Insolation
Inverter
Irradiance
Latitude
Maximum power point tracking (MPPT)
Module
Monocrystalline
Net metering

North American Board of Certified Energy Practitioners (NABCEP)
Off-grid system
Peak sun hours
Photovoltaic (PV) cell
Polycrystalline
Pulse width-modulated (PWM)
Reverse bias
Sea level
Semiconductor
Single-axis tracking
Spectral distribution
Standalone system
Standard Test Conditions (STC)
Sun path
Thin film
Tilt angle
Utility-scale solar generating system
Watt-hours (Wh)

Prerequisites

Before you begin this module, it is recommended that you successfully complete *Core Curriculum*. It is also suggested that you shall have successfully completed the following modules from the Electrical curriculum: *Electrical Level One*, Modules 26101 through 26111; *Electrical Level Two*, Modules 26201, 26205, 26206, and 26208 through 26211; *Electrical Level Three*, Modules 26301 and 26302; and *Electrical Level Four*, Modules 26403 and 26413.

Note: *NFPA 70®*, *National Electrical Code®*, and *NEC®* are registered trademarks of the National Fire Protection Association, Inc., Quincy, MA 02269. All *National Electrical Code®* and *NEC®* references in this module refer to the 2011 edition of the *National Electrical Code®*.

Contents ————————————————————

Topics to be presented in this module include:

Figures and Tables

Figures and Tables (*continued*) ────────

1.0.0 INTRODUCTION

Solar power has been in use for over two thousand years. The ancient Greeks used thick walls to trap heat during the day and release it slowly at night. They also oriented buildings to provide shading in summer while maximizing sunlight in winter. Solar energy was also used to preserve food, heat water, and dry clothes. These early applications of solar power were very effective and are still in use today. They are known as passive forms of solar power because they use sunlight without transforming it.

In 1767, Horace de Saussure experimented with glass boxes to determine how much heat could be trapped by the glass. He discovered that heat could be collected on sunny days even if the outdoor temperature was quite low, as on a mountaintop. These experiments helped scientists understand the effects of atmospheric differences on outdoor temperature and also became the basis for passive solar collectors and solar ovens.

Active forms of solar power did not come into use until 1839, when Alexander Becquerel discovered that certain materials produce electric current when exposed to light. This eventually led to the invention of the first **photovoltaic (PV) cell**. In 1891, Clarence Kemp patented the first water heater powered by solar energy. The first practical PV cell was invented by Bell Laboratories in 1954 using modified silicon cells.

A PV cell consists of two layers of **semiconductor**, one p-type (positive) and the other n-type (negative). The electrical contacts are screened in a grid on both sides of the panel. High-energy light

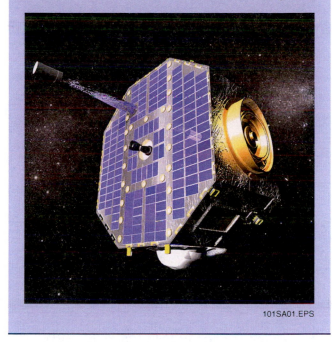

Did You Know?

Solar energy is still the primary source of power in satellite systems and spacecraft. This photo shows the array on a spacecraft designed to map the location of our solar system in the Milky Way galaxy.

101SA01.EPS

particles known as photons strike the semiconductor atoms and release free electrons, producing current (*Figure 1*). Many cells are combined in a solar panel, and these modular panels are connected in an **array**.

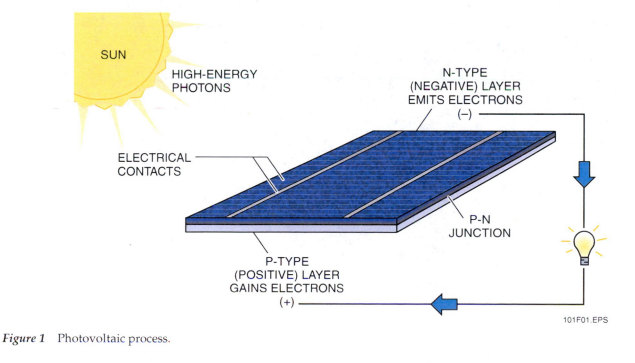

Figure 1   Photovoltaic process.

As America entered the space race in the 1950s, scientists realized that solar arrays provided an ideal power source for satellites. Later, solar cells became popular for operating rural telephones and radio transmitters, and for homes that were too remote to be served by the electrical grid.

In the 1970s, the energy crisis focused more attention on alternative energy sources, and the use of solar power increased in both residential and commercial applications (*Figure 2*). Today, solar power is used for everything from hand-held calculators to giant solar farms that produce enough energy to operate a small city.

PV power has many advantages over other energy sources. Sunlight is a limitless resource, and most areas can support some degree of solar power. PV power is clean and environmentally friendly. It can be harnessed without disrupting the environment and produces no hazardous waste or emissions. It is also quiet, reliable, and requires little maintenance. In addition, PV power can be generated onsite, and does not have the mining/drilling or transportation requirements of fossil fuels. The power generated is also a domestic product, which strengthens the economy, produces jobs, and reduces reliance on foreign energy.

PV systems help to offset power use during peak demand periods. The electrical grid typically reaches peak demand during the late afternoon when people come home from work and turn on the air conditioning. For a solar-powered facility, this is a time when it is producing energy. Putting energy back into the grid during peak demand periods reduces the possibility of **brownouts**. As electricity is produced, it is used at the home or business first, and then any excess electricity is sent into the grid. This is known as **net metering**. Customers receive credits for generating power, which can help to offset the initial cost of installing a PV system. In addition, customers often receive government incentives and utility rebates when installing these systems.

The use of solar power is now increasing by more than 25 percent per year, creating a huge demand for certified installers. This module covers the basic concepts of PV systems and their components. It also explains how PV systems are sized, designed, and installed. Successful completion of this module will help prepare trainees for the **North American Board of Certified Energy Practitioners (NABCEP)** PV Entry Level Exam. Trainees who pass the NABCEP exam may work as helpers installing PV systems. After gaining additional training and logging the required number of installations, trainees will then be qualified to take the NABCEP Certified PV Installer Exam.

101F02.EPS

Figure 2 Typical commercial PV system.

2.0.0 APPLICATIONS

PV arrays can be installed using a wide variety of mounting methods. Some of these include roof mounting, pole mounting, ground mounting, and wall mounting. They can also be installed as part of a shade structure, such as a carport. In addition to mounting methods, PV arrays are also classified by how they are connected to other power sources and loads. PV systems can operate connected to or independent of the utility grid. They can also be connected to other energy sources, such as wind turbines, and energy storage systems, such as batteries. Solar energy can be classified into four basic types of systems: **standalone systems**, **grid-connected systems**, **grid-interactive systems**, and **utility-scale solar generating systems**.

2.1.0 Standalone Systems

Standalone PV systems can be either direct-drive or battery-powered. These systems are commonly used to provide power in areas where access to the grid is inconvenient or unavailable.

Direct-drive systems power DC loads, such as ventilation fans, irrigation pumps, and remote cattle watering systems. Because these systems operate only when the sun is shining, it is essential to match the power output of the array to the load.

Battery-powered standalone systems use PV energy to charge one or more batteries. The battery system then supplies either a DC load or an AC load when an **inverter** is used to convert the DC to AC. Battery-powered standalone systems range from handheld electronics to trailer-mounted systems used to provide emergency power (*Figure 3*). They can also be used for remote data monitoring, emergency highway signage, and street or parking lot lighting (*Figure 4*).

Standalone systems are also used to power buildings in remote areas. These systems do not require power poles and transmission lines, and therefore eliminate the possibility of transmission failures due to downed power lines. These are known as **off-grid systems**. Off-grid systems use batteries for energy storage as well as battery-based inverter systems. **Charge controllers** are used to maximize the battery-charging efficiency of the solar array. Backup power is provided by an engine-driven generator.

101SA02.EPS

101F03.EPS

Figure 3   Portable PV generator.

2.2.0 Grid-Connected Systems

Grid-connected systems operate in parallel with the utility grid. Also known as **grid-tied systems**, these systems are designed to provide supplemental power to the building or residence (*Figure 5*). Since they are tied to the utility, they only operate when grid power is available.

Grid-tied systems invert the DC produced by a solar array into AC, which is then sent to the building's electrical panel to supply power. A DC disconnect is required between the array and the inverter, while an AC disconnect is required between the inverter and the building service panel.

Grid-tied systems require the installation of a special meter by the utility. During the daytime, any power in excess of the load is sold back to the utility in the form of credits. This can be shown by the building's electrical meter, which effectively runs backward when excess energy is supplied. At night and during periods when the load exceeds the system output, the required power is supplied by the electric utility.

101F04.EPS

Figure 4   Solar-powered parking lot fixture.

INVERTER AND DC DISCONNECT

METER

PANEL

(AC DISCONNECT NOT SHOWN.)

101F05.EPS

Figure 5   Grid-tied PV system.

The inverter allows for the conversion of power and also shuts down the system during a power outage or other electrical failure. This is a safety feature that prevents voltage from traveling back into the grid via backfeed.

2.3.0 Grid-Interactive Systems

Like a grid-tied system, a grid-interactive system is connected to the utility and uses inverted PV power as a supplement. Unlike a grid-tied system, which is literally tied to the grid and cannot function independently, a grid-interactive system provides a means of independent power. Grid-interactive systems include batteries that can supply power during outages and after sundown.

When the PV system operates, it first charges the batteries, then it satisfies the existing load, and then any excess power is sent to the grid. As with off-grid systems, a charge controller is used to monitor the battery charge to ensure consistent power with minimal downtime. In the event of a power outage, battery power is supplied to critical loads through the use of a designated subpanel. When these systems are used with other energy sources, such as wind turbines or generators, they are known as hybrid systems.

2.4.0 Utility-Scale Solar Generating Systems

There are two types of utility-scale solar generating systems. The first type is a steam generator that heats the water using concentrated sunlight rather than fossil fuel. This type of generating system is known as a solar boiler (*Figure 6*). The second type uses a large bank of solar cells to produce direct current (*Figure 7*). Motorized systems are often used to adjust the panel position so that it follows the movement of the sun throughout the day. This is known as tracking.

Utility-scale solar generating systems are not yet in common use, but will become increasingly popular as the technology advances. This module focuses on grid-tied and grid-interactive systems used in residential and commercial applications.

3.0.0 OHM'S LAW AND POWER

An understanding of Ohm's law is essential when determining the power and load requirements of PV systems. Ohm's law defines the relationship between current (I), voltage or electromotive force (E), and resistance (R). There are three ways to express Ohm's law mathematically.

- The current in a circuit is equal to the applied voltage divided by the resistance:

$$I = \frac{E}{R}$$

- The resistance of a circuit is equal to the applied voltage divided by the current:

$$R = \frac{E}{I}$$

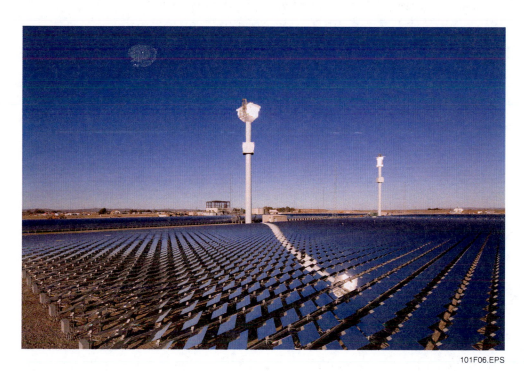

101F06.EPS

Figure 6   Sun-tracking solar boiler.

Introduction to Solar Photovoltaics

Figure 7 Solar power plant.

- The voltage applied to a circuit is equal to the product of the current and the resistance:

$$E = I \times R$$

If any two of the quantities are known, the third can be calculated. The Ohm's law equations can be practiced using an Ohm's law circle, as shown in *Figure 8*. To find the equation for E, I, or R when two quantities are known, cover the unknown third quantity. The other two quantities in the circle will indicate how the covered quantity may be found.

Example 1:

Find I when E = 120V and R = 30 ohms.

$$I = \frac{E}{R}$$

$$I = \frac{120V}{30 \text{ ohms}}$$

$$I = 4A$$

Example 2:

Find R when E = 240V and I = 20A.

$$R = \frac{E}{I}$$

$$R = \frac{240V}{20A}$$

$$R = 12 \text{ ohms}$$

Example 3:

Find E when I = 15A and R = 8 ohms.

$$E = I \times R$$

$$E = 15A \times 8 \text{ ohms}$$

$$E = 120V$$

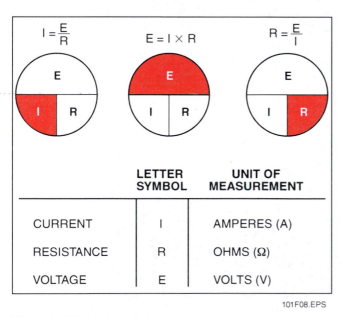

	LETTER SYMBOL	UNIT OF MEASUREMENT
CURRENT	I	AMPERES (A)
RESISTANCE	R	OHMS (Ω)
VOLTAGE	E	VOLTS (V)

Figure 8 Ohm's law circle.

3.1.0 Applying Ohm's Law to Series and Parallel Circuits

You will recall that loads can be arranged in series, in parallel, or in series-parallel. *Figure 9* shows these three types of circuits.

3.1.1 Series Circuits

A series circuit provides only one path for current flow and is a voltage divider. The total resistance of the circuit is equal to the sum of the individual resistances. The 12V series circuit in *Figure 9* has two 30-ohm loads. The total resistance is therefore 60 ohms. The amount of current flowing in the circuit is calculated as follows:

$$I = \frac{E}{R}$$

$$I = \frac{12V}{60\ ohms}$$

$$I = 0.2A$$

The current flow is the same through all the loads. The voltage measured across any load (voltage drop) depends on the resistance of that load. The sum of the voltage drops equals the total voltage applied to the circuit. An important trait of a series circuit is that if the circuit is open at any point, no current will flow.

3.1.2 Parallel Circuits

In a parallel circuit, each load is connected directly to the voltage source; therefore, the voltage drop is the same through all loads and current is divided between the loads. The source sees the circuit as two or more individual circuits containing one load each. In the parallel circuit in *Figure 9*, the source sees three circuits, each containing a 30-ohm load. The current flow through any load is determined by the resistance of that load. Therefore, the total current drawn by the circuit is the sum of the individual currents. The total resistance of a parallel circuit is calculated differently from that of a series circuit. In a parallel circuit, the total resistance is less than the smallest of the individual resistances.

For example, each of the 30-ohm loads draws 0.4A at 12V; therefore, the total current is 1.2A:

$$I = \frac{E}{R}$$

$$I = \frac{12V}{30\ ohms}$$

$$I = 0.4A \times 3 = 1.2A$$

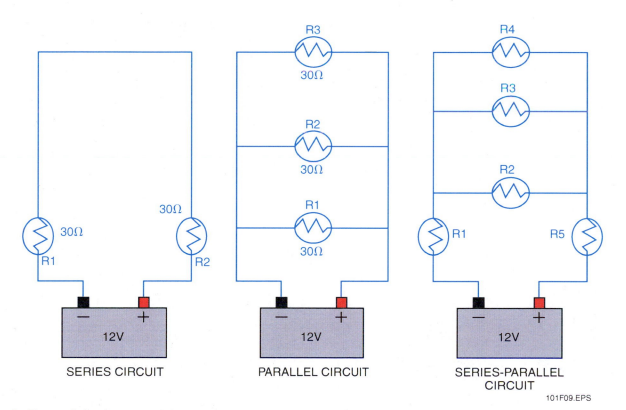

Figure 9 Types of circuits.

Now, Ohm's law can be used again to calculate the total resistance:

$$R = \frac{E}{I}$$

$$R = \frac{12V}{1.2A}$$

$$R = 10 \text{ ohms}$$

3.1.3 Series-Parallel Circuits

When loads are connected in series-parallel, the parallel loads must first be converted to their equivalent resistances. Then the load resistances are added to determine the total circuit resistance.

3.2.0 Ohm's Law and Power

Power is the ability to do work. Mechanical power is often expressed in horsepower (hp). In electrical circuits, power is measured in watts (W). One hp equals 746 watts. One watt is the power used when one ampere of current flows through a potential difference of one volt. For this reason, watts are referred to as volt-amps (VA). Power (P) is determined by multiplying the rated current (I) by the rated voltage (E):

$$P = I \times E$$

This equation is sometimes called Ohm's law for power, because it is similar to Ohm's law. This equation is used to find the power consumed by a circuit or load when the values of current and voltage are known. Using variations of this equation, the power, voltage, or current in a circuit can be calculated whenever any two of the values are known. See *Figure 10*. Note that all of these formulas are based on Ohm's law ($E = I \times R$) and the power formula ($P = I \times E$).

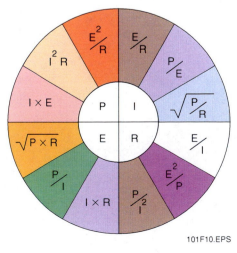

Figure 10 Expanded Ohm's law circle.

Watt-hours (Wh) are calculated by multiplying the power in watts (VA) by the number of hours during which the power is used. The kilowatt-hour (kWh) is commonly used for larger amounts of electrical work or energy. (The prefix *kilo* means one thousand.) For example, if a light bulb uses 100W or 0.1kW for 10 hours, the amount of energy consumed is 0.1kW × 10 hours = 1.0kWh.

Very large amounts of electrical work or energy are measured in megawatts (MW). (The prefix *mega* means one million.)

3.3.0 Series and Parallel Circuits in Solar PV Systems

Household circuits and most other loads are wired in parallel. However, solar panels can be wired in series, in parallel, or in series-parallel, depending on the desired output. Wiring solar panels in series increases the voltage, while wiring them in parallel increases the amperage.

When solar panels are wired in series, the positive terminal of one panel is connected to the negative terminal of another. Unlike loads that drop voltage and use power, when you put solar panels in series they are creating power. This doubles the voltage, but the total amperage remains the same. For example, a 36V/5A panel delivers 180W of power (VA). Two 36V/5A panels wired in series will produce 72V at 5A or 360W. Four 36V/5A panels wired in series will produce 144V at 5A or 720W.

When two solar panels are wired in parallel, the positive terminal of one panel is connected to the positive terminal of the next panel. This doubles the amperage, but the voltage stays the same. For example, two 36V/5A panels wired in parallel will produce 36V at 10A. Four 36V/5A panels wired in parallel will produce 36V at 20A.

Solar arrays can be connected in series-parallel to achieve the desired voltage and current of the total array (*Figure 11*). For example, two 36V/5A panels can be wired in series to produce 72V at 5A, and then connected to two panels wired in parallel to produce a total output of 72V at 10A. If these four panels are then parallel-connected to four other panels wired in the same way, the voltage will be 72V at 20A, and so on. This is the basic concept behind building solar arrays.

3.4.0 Peak Sun and Power

To determine the total panel or array output, the wattage is multiplied by the **peak sun hours** per day for the geographical area to determine the watt-hours produced per day. Peak sun hours or **insolation** values represent the equivalent number of hours per day when solar **irradiance** averages 1,000W/m². Irradiance is a measure of radiation

NCCER — *Contren® Learning Series* 57101-11

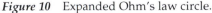

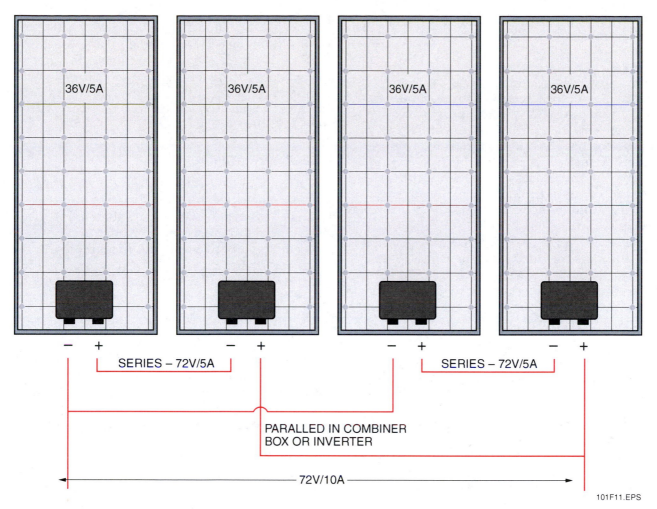

Figure 11 Series-parallel wiring in a solar array.

density and varies widely by location. Areas with few cloudy days and low levels of dust have high levels of irradiance. In addition, higher **elevations** have greater irradiance than those at **sea level**. Higher elevations also have cooler temperatures, which reduces the resistance and increases the array output voltage.

The peak sun hours for a given location represent an average value since solar intensity varies by time of day, season, and cloud cover. For example, a location may receive 800W/m² for three hours and 1,200W/m² for two hours. The total would be 4,800W/m² divided by 1,000W/m² = 4.8 peak sun hours.

To convert peak sun hours to power, multiply the wattage by the peak sun hours. For example, the power produced by a 36V/5A solar panel is 36V × 5A = 180W. If this 180W panel receives five peak sun hours per day, it will produce 180W × 5 = 900 watt-hours (0.9kWh) per day.

> **NOTE**
> Insolation maps and other resources can be found on the National Renewable Energy Laboratory (NREL) website at www.nrel.gov.

4.0.0 PV System Components

A typical grid-interactive PV system has four main components: the panels, an inverter, batteries, and a charge controller. The remaining components are known as the **balance-of-system (BOS)** components. They include the panel mounts, wiring, overcurrent protection, grounding system, and disconnects. PV installations must comply with all applicable requirements of the *National Electrical Code® (NEC®)*. ***NEC Article 690***

contains specific requirements for the installation of PV systems.

4.1.0 PV Panels

PV panels, sometimes referred to as **modules**, consist of numerous cells sealed in a protective laminate, such as glass. Solar cells work using a semiconductor that has been **doped** to produce two different regions separated by a p-n junction. Doping is the process by which impurities are introduced to produce a positive or negative charge. Crystalline silicon (c-Si) is used as the semiconductor in most solar cells.

PV panels are normally prewired and include positive and negative leads attached to a sealed termination box. Panels are rated according to their maximum DC power output in watts under **Standard Test Conditions (STC)**. Standard Test Conditions are as follows:

- Operating temperature of 25°C (77°F)
- Incident solar irradiance level of 1,000W/m²
- **Air mass (AM)** of less than 1.5 **spectral distribution**

Spectral distribution is caused by the distortion of light through Earth's atmosphere. The air mass value is based on an ideal value of zero, which occurs in outer space.

Because panels are rated under ideal conditions, their actual performance is usually 85 to 90 percent of the STC rating. The output of a PV cell depends on its efficiency, cleanliness, orientation, amount of sunlight, and temperature. Temperature increases can cause a significant decrease in the system output voltage. This is because higher air temperatures decrease irradiance. The irradiance drops by 50 percent for every 10°C temperature rise (18°F). (Note that this isn't a straight conversion because it represents an interval, rather than a defined temperature.)

PV cells are characterized by the type of crystal used in them. There are three basic types of PV cells: **monocrystalline**, **polycrystalline**, and **amorphous** (commonly known as **thin film**). The type of crystal used determines the efficiency of the cell. For example, a typical monocrystalline panel might produce 75W of power, while an equivalent polycrystalline cell might produce 65W, and a thin-film cell might produce 45W.

> **NOTE**
> Actual panel wattages vary widely depending on manufacturer, panel size, and type of semiconductor. Always consult the nameplate data for the specific panel in use.

4.1.1 Monocrystalline

Monocrystalline cells are formed using thin slices of a single crystal. Monocrystalline cells are currently the most efficient type of PV cell. Due to the manufacturing process, however, they are also the most expensive. A typical monocrystalline panel is shown in *Figure 12*.

4.1.2 Polycrystalline

Polycrystalline cells are made by pouring liquid silicon into blocks and then slicing it into wafers. This is a less expensive process than using a single crystal. However, poured silicon creates non-uniform crystals when it solidifies, reducing the efficiency of the panel. This gives these panels their characteristic flaked appearance (*Figure 13*).

4.1.3 Thin Film

Thin-film PV panels are made using ultra-thin layers of semiconductor material. The reduced material use results in lower manufacturing costs, but also produces the lowest efficiency. Because of this, solar panels using thin-film cells must be larger to produce the same amount of energy as the other types.

Thin-film cells are often used in low-voltage applications, such as solar calculators and other electronics. They can also be encased in laminate to create rigid solar panels or used to create flexible building materials, such as for alternate rows

101F13.EPS

Figure 12 Monocrystalline PV panel.

101F13.EPS

Figure 13   Polycrystalline PV panel.

in roofing shingles (*Figure 14*). When a solar panel is built into a structure, it is known as **building-integrated photovoltaics (BIPV)**.

4.2.0 Inverters

Inverters are used to convert the DC produced by the PV array into AC that can be used by various loads (*Figure 15*). As you will recall, AC travels in a sine wave. There are two common types of inverters: modified sine wave and true sine wave. Modified sine wave inverters are less expensive but do not provide the power quality of true sine wave inverters. They are not recommended for use with electronic equipment or other sensitive devices, including certain types of motors. True sine wave inverters produce smooth power similar to that provided by the grid. The quality of this power may actually be superior to grid power as it is free of voltage dips, spikes, and noise.

Inverters are available in a wide variety of types and load ratings. Some inverters are designed to operate with standalone systems, while others are designed for grid-tied and grid-interactive systems. Off-grid systems use battery-based inverter systems with an output voltage of 120/240VAC so they can operate larger loads such as water heaters and stoves. Inverters are rated in both continuous watts and surge watts. Some inverters include integral ground-fault protection and DC disconnects, while others require that these devices be installed separately. Many inverters use a digital display to indicate **ambient temperature**, voltage output, and other system parameters. The display is normally in standby mode and appears blank. To access various system outputs, simply tap on the inverter housing and the values are displayed.

4.3.0 Batteries

Batteries are used to store energy produced by the PV array and supply it to electrical loads as needed (*Figure 16*). Batteries are also used to operate the PV array near its maximum power point, to stabilize output voltages during periods of low production, and to supply startup currents to loads such as motors. All stationary battery installations must comply with the provisions of *NEC Article 480*.

THIN-FILM SOLAR PANEL

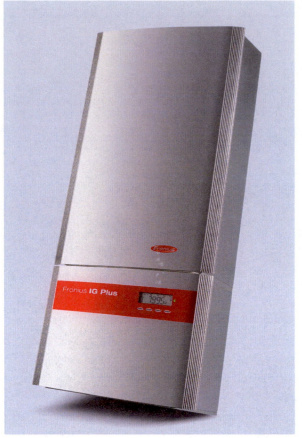

101F15.EPS

Figure 15   True sine wave inverter.

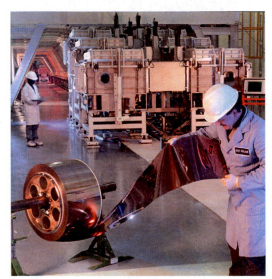

FLEXIBLE THIN FILM FOR BIPV APPLICATIONS

101F14.EPS

Figure 14   Typical thin-film applications.

Solar batteries require proper charging. While some deep cycle batteries can be discharged to 20 percent of capacity, it is best not to go that low as repeated deep cycling shortens battery life. The amount of charge remaining is known as the **depth of discharge (DOD)**. Battery capacity is measured in amp-hours (Ah). For example, a 100Ah battery from which 80Ah has been withdrawn has undergone an 80 percent DOD. Some PV systems include a low battery warning light or cutoff switch to prevent the damage caused by repeated deep cycling.

101F16.EPS

Figure 16   12V PV Batteries.

Batteries must be protected from extreme temperatures. Cold temperatures reduce battery output while hot temperatures increase deterioration. With proper care and appropriate cycling, solar batteries can last between 5 and 7 years. Batteries are rated by the expected number of charge cycles as well as the maximum discharge current. Each cell produces 2V, so a 6V battery contains three cells. Batteries can be connected in series or in parallel to produce the desired output. Residential PV systems are limited to no more than 24 two-volt cells connected in series (48V total) per *NEC Section 690.71(B)(1)*.

> **NOTE**
>
> There are stringent code requirements for battery banks in excess of 48V. See *NEC Section 690.71* for details.

There are two types of solar batteries: flooded lead acid (FLA) and sealed absorbent glass mat (AGM).

4.3.1 FLA Batteries

FLA batteries are more common and less expensive than AGM batteries. However, they require regular maintenance to add water and prevent corrosion. FLA batteries can deteriorate quickly if the fluid levels drop and the plates oxidize. They require good air circulation to prevent gases from accumulating to explosive levels or causing corrosion of other equipment. They are normally located in a separate compartment with appropriate venting and warning labels.

> **WARNING!**
>
> Overcharging and improper venting can cause an explosive gas buildup within a battery. When this occurs, the battery may leak, feel warm, or exhibit signs of swelling. The pressure rises within the cells until a nearby spark or short circuit ignites the gas. Battery explosions release violent sprays of caustic acid and shrapnel that can cause severe injury or even death.

4.3.2 AGM Batteries

AGM batteries are sealed lead-acid batteries. The acid is contained in a special glass fiber mat so a vented battery compartment is not required. In addition, AGM batteries generally do not leak or produce corrosive gases. These batteries are also maintenance-free.

4.4.0 Charge Controllers

Charge controllers are used to regulate the charge and discharge of the system batteries. They are rated by their maximum AC current output, so it essential that they be sized to match the application. There are two main types of charge controllers: **pulse width-modulated (PWM)** and **maximum power point tracking (MPPT)**. Older charge controllers use shunts or relay transistors and do not work well with sealed batteries. PWM charge controllers are ideal for use with sealed batteries but can only be used within a limited range of panel configurations and voltages. An MPPT charge controller is shown in *Figure 17*. These controllers harvest energy more efficiently and can be used with a wider range of array configurations and voltages. A digital display is provided to monitor power use, battery charge, and other system performance data. MPPT charge controllers are preferred for use in cold climates because they prevent the overvoltage that can occur at lower temperatures and provide more precise control over the DOD.

4.5.0 BOS Components

BOS components include the wiring, grounding system, disconnects, foundations, and support frames. They also include any required subpanels,

101F17.EPS

Figure 17   MPPT charge controller.

conduit, and combiner boxes. Weatherproof combiner boxes are used to connect strings of solar panels to create a larger array, and to provide a convenient array disconnect point.

4.5.1 Electrical System Components

A PV electrical system includes all wiring, devices, and components between the service entrance and the final panel termination. The most important safety devices in a PV system are the AC and DC disconnects and the grounding system.

Each piece of equipment in a PV system requires a switch or breaker to disconnect it from all sources of power per *NEC Section 690.15.* Disconnect requirements are listed in *NEC Section 690.17.* A DC disconnect is shown in *Figure 18.* This disconnect is attached to the inverter. If the inverter does not include a DC disconnect, it must be purchased and installed separately.

The AC disconnect is located near the service panel. A typical AC disconnect is shown in *Figure 19.* PV disconnects may be located indoors if the panel is in the building.

> **NOTE**
>
> PV disconnects are not allowed in residential bathrooms per *NEC Section 690.14(C)(1).*

The grounding system is essential to the electrical integrity of a PV array. All exposed non–current-carrying metal parts of the equipment must be bonded and connected to ground. The ground connection must be continuous. The roof ground should be a minimum 6 AWG solid copper to provide extra strength. Always use approved cleats when making ground connections. They are designed to pierce the finish of the panel to

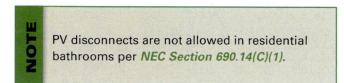

101F18.EPS

Figure 18   DC disconnect.

101F19.EPS

Figure 19   AC disconnect.

ensure electrical continuity. A ground rod connection is shown in *Figure 20.*

Many ground-mounted PV systems incorporate multiple levels of protection against stray voltages, such as lightning. These are installed in addition to the equipment grounding system and may include surge protectors and the use of grounded metal fencing for ground-mounted arrays.

Ground-fault protection is required per *NEC Section 690.5.* Some inverters include integral ground-fault protection. If it is not supplied in the inverter, it must be located elsewhere in the PV system.

Other electrical system BOS components include the panel wiring, conduit, combiner boxes, and termination boxes. See *Figure 21.* All wiring must be sized for the load and the distance between components. All cables carrying other-than-low voltages must be routed in conduit.

4.5.2 Footers and Support Structures

Support frames and footers are the most labor-intensive portion of any ground-mounted installation. In a system installed in an open field, the

Figure 20 Grounding system.

Figure 21 BOS components.

Figure 22 Roof-mounted rail support.

footers are typically concrete while the support frames are aluminum or steel.

Roof-mounted systems are mounted on aluminum or stainless steel rails secured into the roof trusses using lag bolts. See *Figure 22*. These supports include integral flashing to prevent water entry. Proper weathersealing of all roof penetrations is essential.

5.0.0 SAFETY CONSIDERATIONS IN PV SYSTEMS

PV system installation often involves working in extreme heat on slanted, slippery roof surfaces ten or more feet above the ground (*Figure 23*). Always wear sunscreen, sunglasses, and non-slip footwear. Use fall protection in the form of guardrails. If guardrails cannot be installed, use a personal fall arrest system (PFAS). Ensure that straight ladders are positioned at an angle of one-fourth of the working height and extend beyond the roof surface by 36 inches. Maintain control over all

tools, equipment, and system components so they do not fall or slide off the roof, possibly injuring workers below. Wear a hardhat at all times. Avoid working in wet conditions. Never work where you can see lightning or hear thunder. Be careful when handling panels and other flat materials in windy conditions.

Never look directly into the sun. Avoid excessive sun exposure and stay hydrated. Drink eight ounces of water for every fifteen minutes spent working in the heat, as failure to stay hydrated can result in heat exhaustion or heat stroke. Do not use caffeinated drinks or alcohol before or during work.

In addition to working at elevation, you will be handling large, sharp, unwieldy PV panels that have the capacity to become energized the moment light strikes them (*Figure 24*). Plug the leads together to avoid the possibility of shock. Keep panels covered per the manufacturer's instructions so they remain de-energized. Wear protective gloves and seek assistance when lifting to avoid back strain.

PV panels may be hoisted onto a scaffold or moved using a material handling system. Scaffolds may only be erected and inspected by qualified individuals. All scaffold workers must be trained in the safe use of this equipment. Follow all safety procedures when using mechanical moving equipment. Never operate any equipment unless you are fully qualified.

Working with PV systems also involves working in hot attic areas. Provide ventilation where possible and avoid contact with insulation. Identify

Figure 23 PV array installed on a terra cotta roof.

Figure 24 Installing PV panels.

exit routes before working in tight areas. Exercise caution when working in attic spaces to avoid falling through the ceiling below. Install wood across ceiling joists to provide secure footing.

5.1.0 Fall Protection

The most common safety hazard when installing PV systems on an elevated structure is falls. This is compounded by the fact that there is usually no overhead attachment point, which is the approved method for protection. OSHA standards require PFAS to control falls for any working surface six feet or more above the ground. If you are forced to attach below the D-ring on your harness, you will fall that distance plus the six-foot length of a standard lanyard. This could result in a fall of up to 12 feet. A PFAS is not designed to meet those forces and will transfer much higher impacts when it arrests the fall. To combat this, the American National Standards Institute (ANSI) has changed the equipment manufacturing standards for fall protection. The new standard is commonly referred to as the 12-foot lanyard. The lanyard is still only six feet long, but the allowable deployment (extendable length) of the shock absorber has been increased. This lessens the force transferred to the body through the harness, reducing the likelihood of injury.

Before using a PFAS, you must examine the space below any potential fall point. Make sure it is clear of any hazards or obstructions. When assessing the required free length of a PFAS, you must account for the deployment of the shock absorber. In addition, PFAS manufacturers require up to a three-foot safety factor. Taking this safety factor into account, *Table 1* indicates the minimum distance required between any hazard/obstruction and the worker or attachment point.

As you can see, a typical single-story building may not provide enough height to use standard lanyards. Retractable lanyards that engage and stop falls in less than two feet are usually the best choice for fall protection when used below 20 feet or on any type of lift equipment.

ANSI has also addressed the locking mechanism (gate) on snap hooks. OSHA requires the gate to be rated for 350 pounds when the rest of the system is rated at 5,000 pounds. This creates a weak link in the PFAS. ANSI issued a new standard increasing the strength of these gates to 3,600 pounds.

> **NOTE**
>
> Ladders, scaffolds, stairs, and similar types of equipment have other fall protection standards. Make sure you know and follow your company safety policies when using this equipment.

5.2.0 Battery Hazards

Deep cycle batteries release hydrogen when they are overcharged. This gas can build up to unsafe levels if batteries are poorly vented. The pressure rises within the cells until a spark or short circuit ignites the gas and it explodes. Battery explosions can cause severe injury or death. Replace any

On Site

Battery Temperature

Batteries will begin to show increased temperatures prior to failure. A battery temperature sensor can be used with certain charge controllers to monitor the batteries and shut down the system at excessive temperatures.

101SA03.EPS

Table 1 Required Free Lengths for Various Lanyards

Lanyard Type	Lanyard Length	Shock Absorber Length	Average Worker Height	Safety Factor	Required Free Length
Standard 6' lanyard with 3.5' shock absorber	6'	3.5'	6'	3'	18.5'
New ANSI 6' lanyard with 4' shock absorber	6'	4'	6'	3'	19'
New ANSI 12' lanyard with 5' shock absorber	6' (12' freefall)	5'	N/A (accounted for by lanyard)	3'	20'

battery that feels unusually warm, appears swollen, or leaks. When handling batteries, wear approved, chemical-rated safety equipment including eye and face protection, gloves, and aprons.

Placing a wire across the terminals of a battery is called shorting the battery. Some batteries will explode when shorted. Never short a battery.

5.3.0 Electrical Hazards

Before working on any electrical equipment, the equipment and all inputs/outputs must be placed into an electrically safe work condition. This means that all equipment must be de-energized, any stored electrical energy must be discharged, and the equipment must be locked out and tagged before any work can be performed. It must also be tested to ensure that it is in an electrically safe condition using a known working tester.

Remember that PV systems have multiple power sources. These include utility power, PV system power, and batteries (if used). All possible power sources must be locked out and tagged before working on the system.

PV panels are rated up to 600V. In actual use, a PV panel may experience conditions that result in more current and/or voltage than listed on the panel nameplate.

A PV panel generates electricity when exposed to sunlight. Keep the panel covered until ready for installation and use insulated tools. When the panel is removed from the shipping container, plug the leads together to avoid the possibility of shock. Never touch the terminals when handling a panel.

Some of the most common electrical hazards in PV systems include cracked or broken panels, faulty connections, broken ground wires, and undersized wiring. Be aware of all potential hazards and exercise caution during installation and troubleshooting.

5.4.0 Meter Safety

Before using any meter, ensure that it is rated for the given application. PV installations require the use of Category III/IV meters. These meters are rated for use at the higher voltages expected in these systems. Inspect the meter and leads for damage before use. Never use a damaged meter. Wear safety glasses when using meters. Zero the meter before use, then set the meter to the correct range and select AC or DC.

> **NOTE**
> Detailed safety procedures were covered in the *Electrical Safety* module earlier in your training. Review this material.

Backfeed

Solar PV systems may present a backfeed hazard when the grid is de-energized. Do not rely on the inverter to disconnect power when the grid is de-energized. Provide warning labels on all service equipment and always perform lockout/tagout procedures.

101SA04.EPS

6.0.0 SITE ASSESSMENT

A careful site assessment is critical to a successful PV system. It provides an opportunity to interview the customer regarding their budget and expectations. It gives you a chance to review the customer's current electrical loads and the potential energy savings provided by PV power. You will also assess the site for optimal array orientation and equipment location, as well as any potential problems, including roof condition and shading.

> **NOTE**
> Local building codes and/or homeowner associations may place limits on array installation. For example, some local codes require that PV systems be installed on light-colored roofs to maximize light reflection, while some homeowner associations prohibit array installation on the street-facing roof face. Always consult local codes and homeowner association agreements prior to completing a site assessment.

Before leaving to perform a site assessment, ensure that you have the essential tools of the trade. This includes all required personal protective equipment, including sunscreen, protective eyewear, fall protection, non-slip shoes, and a hardhat. You will also need a site survey checklist, camera, calculator, angle finder, compass, ladder, line, and Solar Pathfinder™ (or equivalent).

> **WARNING!**
> When performing a site assessment, always wear sun protection, protective sunglasses, and shoes with non-slip soles. Use the appropriate fall protection and ensure that straight ladders extend beyond the roof surface by 36". To avoid permanent eye damage, never look directly into the sun.

6.1.0 Customer Interview

The customer interview lays the foundation for the system design. To the customer, the primary consideration in system selection is most likely the cost. For example, an entry-level system may supply 640W, which is enough power to run energy-efficient lights, a small refrigerator, and a few small appliances when supplied with batteries to overcome the motor starting currents. In comparison, a 3,600W system may supply enough power to operate a small home, including some larger motor loads. However, it is likely to cost three times as much as an entry-level system. When discussing system expectations with a customer, it is important to determine both their power expectations and budget. Find out whether they are seeking 100 percent replacement for grid power or if they will be satisfied with a partial replacement instead. Determine if the customer requires battery backup for the system.

6.2.0 Power Consumption

The customer interview must also include a load estimate. Many utilities provide free energy audits that can help in determining usage patterns. At a minimum, have the customer provide recent utility bills. In addition to precise kWh use for the current period, many utility bills show energy use over the last twelve months in the form of a simple bar chart on the bill. Customers may also be able to access this data online at their utility provider's website. Whenever possible, have the customer collect this data prior to the site visit. Find the month showing the highest power use, and then divide this value by the number of days in that month to find the average daily load. Multiply by 1,000 to convert kWh to Wh. Adjust this amount

to reflect the customer's desired PV replacement percentage for the grid-connected load.

If current usage information is unavailable, it must be calculated using equipment wattages and expected hours of use. List all the loads to be supplied by the PV system and note the daily hours of use for each item during the critical design month (the coldest month). Multiply this by the nameplate wattage of the equipment to determine the daily Wh requirements.

> **NOTE**
> Appliance wattages can usually be determined from the owner's manual or equipment nameplate. If an appliance is rated in amps, multiply the amperage by the operating voltage to determine the value in watts.

6.3.0 Roof Evaluation

Document the type, size, age, and condition of the roof, including the strength and accessibility of supporting trusses. Consider the potential for wind loads and severe weather. If a roof is likely to require replacement in the next ten years, it should be replaced before the installation of a PV system.

> **WARNING!**
> If the roof structure is not sound, inform the building owner. Either a different location must be considered or the installation must be delayed until the roof is repaired. Never attempt to install a PV array on an unsound structure.

Consider the potential load in reference to the available space for the array. If the roof space is inadequate, a ground-mounted system must be considered. For example, suppose you have a south-facing, obstruction-free roof area of 200 square feet and will be installing polycrystalline PV panels. These panels have an average output of 10 watts per square foot. Multiply the available roof area by the wattage per square foot to find the largest system that can be installed:

200 sq ft × 10W/sq ft = 2,000W or 2.0kWh

This roof is large enough for a 2.0kWh system using typical polycrystalline panels. Note that actual panel outputs can vary widely and this is a general sizing strategy that can be applied to determine whether the roof has enough space for the installation of an array.

Sketch the south-facing roof area, including gables, vents, and other protrusions. Any shading caused by these protrusions will severely reduce the output of the PV array. If protrusions are noted, document an alternative location for the array. Next, determine the roof slope using an angle finder. Photograph the roof and the existing service.

> **NOTE**
> If the roof has terra cotta tile, the array installation must be performed by a specialized roofing contractor.

6.4.0 Array Orientation

Solar arrays work best when facing true solar south. True solar south is slightly different than a magnetic reference or compass south. A quick way to determine true solar south is to measure the length of time between sunrise and sunset, and then divide by two. The position of the sun at the resulting time represents true solar south.

> **NOTE**
> Solar south can also be determined using the National Oceanic and Atmospheric Administration Sunrise/Sunset Calculator at www.srrb.noaa.gov.

PV systems are designed to maximize output by optimal placement in relation to the motion of the sun. The most important values include **azimuth**, **altitude**, and **declination**.

- *Azimuth* – For a fixed PV array, the azimuth angle is the angle clockwise from true north that the PV array faces. The default azimuth angle is 180° (south-facing) for locations in the northern hemisphere and 0° (north-facing) for locations in the southern hemisphere. This value maximizes energy production. In the northern hemisphere, increasing the azimuth favors afternoon energy production, while decreasing it favors morning energy production. The opposite is true for the southern hemisphere.
- *Altitude* – The altitude is the angle at which the sun is hitting the array. In many areas, panels are simply set to an angle that matches the local **latitude**. This provides a yearly average maximum output power. However, seasonal adjustments can increase the output in locations with high irradiance. In small systems, panels can be manually adjusted to optimize the **tilt angle**. Large systems are more likely to use automatic

tilt adjustment. General recommendations for tilt angle include the following:
- *Maximum output power in winter* – In the northern hemisphere during the winter

On Site

Measuring Irradiation

Pyranometers are used to measure irradiance across the entire sky. Some units are mounted at the same angle as the array. They provide long-term data that can be compared against the array output. Other pyranometers are handheld and plug into a digital reader. Handheld units are less precise but offer the advantages of portability and rapid field measurements.

MOUNTED PYRANOMETER

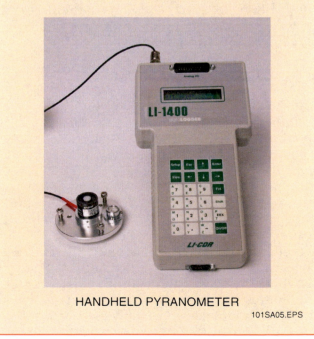

HANDHELD PYRANOMETER

101SA05.EPS

months, the tilt angle should equal local latitude +15°.

- *Maximum output power in summer* – In the northern hemisphere during the summer months, the tilt angle should equal local latitude –15°.

- *Seasonal adjustment* – In March and September, the tilt angle equals the local latitude. In June, the tilt angle equals latitude –15°, and in December, the tilt angle equals latitude +15°. These adjustments ensure maximum efficiency at all times.

• *Declination* – The declination of the sun is the angle between the equator and the rays of the sun. It ranges between +23.45° on the summer solstice (on or about June 21) to –23.45° on the winter solstice (on or about December 21). During the spring equinox (on or about March 21) and the fall equinox (on or about September 21), the angle is zero. Because declination changes throughout the year, the optimal panel tilt angle also changes. This seasonal tilt of the Earth causes longer shadows in winter because the sun is lower in the sky.

The Solar Pathfinder™ is a manual tool used to provide a full year of solar data for a specific location (*Figure 25*). It includes sun path diagrams for various latitude bands and an angle estimator for determining the sun's altitude and azimuth at various times of year. To use this instrument, go to the National Geophysical Data Center website at www.ngdc.noaa.gov and follow the links to find the latitude and declination angle for the desired location. Use the latitude to select the correct band diagram and a compass to align the unit to the desired angle of declination. Next, document shading by tracing the trees in the reflection using a grease pencil. You can also take a picture of the reflection rather than tracing it. The PV system must be installed outside of the shaded area or trees must be removed to eliminate shading. Companion software is available that generates monthly sun paths for each specific site latitude instead of using the latitude band diagrams.

The Solmetric SunEye™ is an electronic device that allows users to instantly assess the total potential solar energy of a site (*Figure 26*). This device shows site shading in a digital display rather than through tracings. It is more expensive than the Solar Pathfinder™ but has the advantage in ease of use.

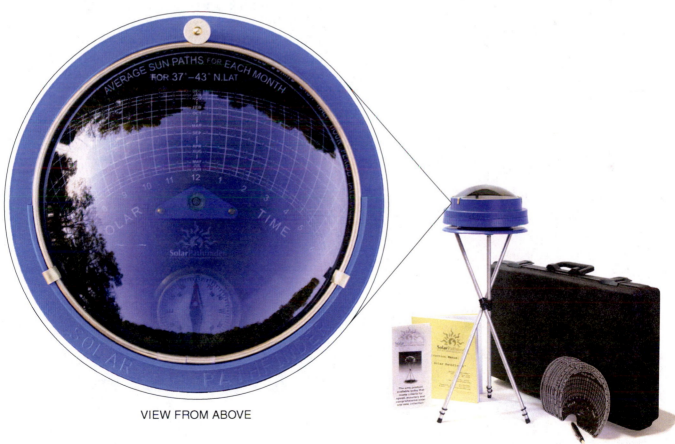

VIEW FROM ABOVE

101F25.EPS

Figure 25   Solar Pathfinder™.

In most locations in the United States, winter produces the least amount of sunlight due to shorter days, increased cloud cover, and the sun's lower position in the sky. Insolation is usually highest in June and July and lowest in December and January. When selecting a site, choose an area that is free of shade between 9 AM and 3 PM on December 21st. *Figure 27* shows the path of the sun during the year for a specific location.

For example, to find the approximate altitude and azimuth of the sun in April, follow along on the curve that reads "Apr. & Aug. 21" until you find 2:30 PM. The altitude is 45° while the azimuth is 60°. Note that values outside of the lines must be estimated.

6.5.0 Equipment Location

Sketch the location of the service in relation to the proposed array. Determine a location for the inverter (near the panel), charge controller (if a battery system will be installed), AC disconnect, DC disconnect, and ground. Group these items together whenever possible. Document a suggested battery location. The batteries should be located close to

101F26.EPS

Figure 26 Solmetric SunEye™.

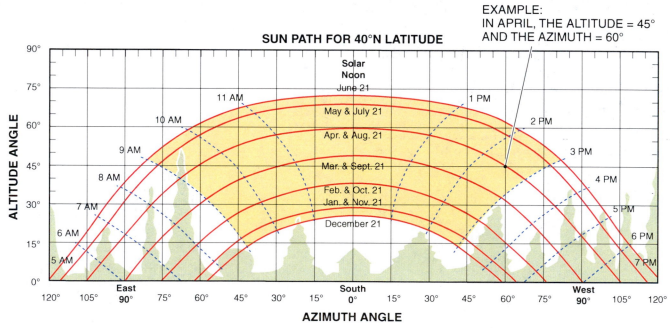

EXAMPLE:
IN APRIL, THE ALTITUDE = 45°
AND THE AZIMUTH = 60°

SUN PATH FOR 40°N LATITUDE

To use this chart for southern latitudes, reverse the horizontal axis (east/west & AM/PM).

101F27.EPS

Figure 27 Sun path diagram.

the array to minimize voltage losses, but must be protected against temperature extremes. FLA batteries must be located in a vented compartment.

A typical site survey checklist is included in the *Appendix*.

7.0.0 SYSTEM DESIGN

PV system design begins with the load data collected during the site assessment. Next, various system choices must be made, such as sizing the PV panels, inverter, batteries, charge controller, and BOS components. All components must be properly matched to ensure efficient output.

7.1.0 Panel Nameplate Data

The panel nameplate contains a wealth of information that can be used in system sizing and design. It lists the model number, power output (P_{MAX}), maximum voltage and current at P_{MAX}, short-circuit current (I_{SC}), and open-circuit voltage (V_{OC}). The I_{SC} and V_{OC} are critical values when performing system calculations. The nameplate also lists the normal operating cell temperature (NOCT) at STC conditions, panel weight, maximum system voltage, fuse rating, and recommended conductor. See *Figure 28*. The minimum wire size listed on this panel nameplate is 12 AWG rated for 90°C. Type USE-2 cable is commonly used.

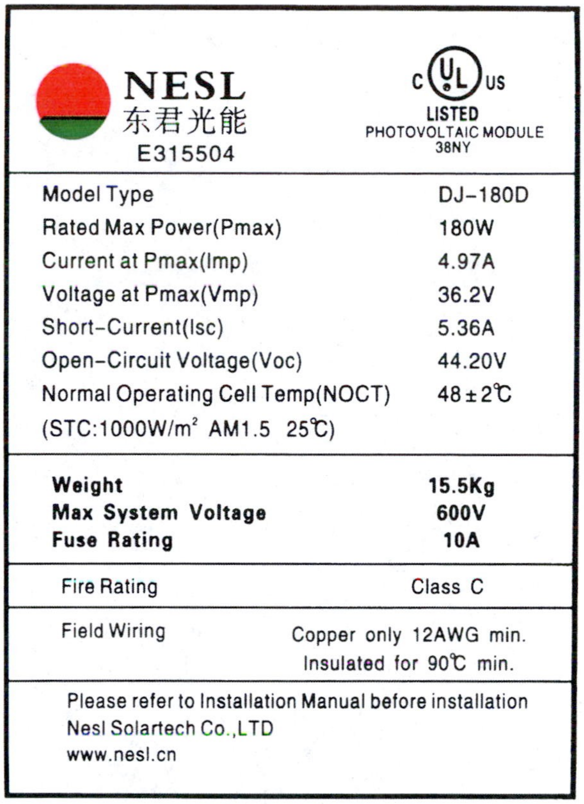

> **NOTE**
> The panel shown in *Figure 28* has a Class C Fire Rating and must be installed over a roof of equivalent fire resistance.

7.2.0 Solar Array Sizing

PV system sizing starts with the load and works backwards. To find the correct array size, divide the daily load in watt-hours determined in the site assessment by the number of sun hours per day, adjusted for the desired replacement percentage of grid power. For example, suppose a customer lives in an area with four peak sun hours per day, uses an average of 30kWh per day (30,000Wh/day) during the critical design month, and desires 50 percent replacement of grid power. The size of this PV array is calculated as follows:

$$30,000\text{Wh/day} \div 4 \text{ peak sun hours/day} =$$
$$7,500\text{Wh/day} \times 0.5.0 = 3,750\text{Wh/day}$$

The panel's I_{SC} rating must be multiplied by a factor of 1.25 when determining component ratings. *NEC Section 690.8* requires an additional multiplying factor of 1.25 for conductor and fuse

NESL
东君光能
E315504

c(UL)us
LISTED
PHOTOVOLTAIC MODULE
38NY

Model Type	DJ-180D
Rated Max Power(Pmax)	180W
Current at Pmax(Imp)	4.97A
Voltage at Pmax(Vmp)	36.2V
Short–Current(Isc)	5.36A
Open–Circuit Voltage(Voc)	44.20V
Normal Operating Cell Temp(NOCT)	48±2℃
(STC:1000W/m² AM1.5 25℃)	

Weight	**15.5Kg**
Max System Voltage	**600V**
Fuse Rating	**10A**

Fire Rating	Class C

Field Wiring	Copper only 12AWG min. Insulated for 90℃ min.

Please refer to Installation Manual before installation
Nesl Solartech Co.,LTD
www.nesl.cn

101F28.EPS

Figure 28   PV nameplate data.

sizing. Refer to *NEC Table 690.7* for voltage correction factors based on ambient air temperatures.

7.3.0 Inverter Selection

Inverters are rated for specific battery voltages as well as for continuous wattage and surge wattage. The continuous wattage represents the total power output the inverter can support over time. The surge wattage accounts for momentary inrush currents. To select the appropriate inverter size, add up the wattages of all loads that are likely to operate at the same time. This determines the minimum continuous wattage. Next, consider the potential surge of each load to determine the minimum surge wattage. This is typically 150 to 200 percent of the continuous rating. For example, a 1,500W inverter might have a 2,500W surge rating.

Inverters are also selected based on the application. Off-grid systems use inverters with integral battery backup, while grid-tied systems use standard inverters (*Figure 29*).

All inverters list their efficiency ratings in the product data. For example, a typical grid-tied 2,000W inverter might have an efficiency of 95 percent. Values near or slightly above 95 percent are common. The inverter efficiency must be taken into consideration when sizing the inverter for the application.

STANDARD THREE-PHASE INVERTER

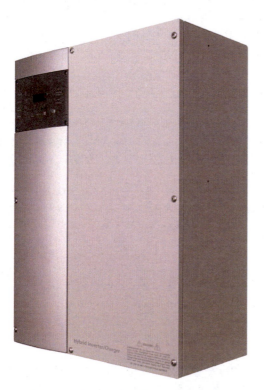

INVERTER WITH BATTERY SYSTEM

101F29.EPS

Figure 29 Inverters are selected based on the desired size and application.

7.4.0 Battery Bank Sizing

The size of the battery bank depends on the storage capacity required, the maximum charge and discharge rates, and the minimum temperature at which the batteries will be used. Wiring batteries in series increases the voltage. Wiring batteries in parallel increases the current (amp-hour) capacity, but does not affect the voltage. *Figure 30* shows a typical battery installation.

The amp-hour requirements of DC batteries must account for the voltage difference between the system and the batteries. For example, suppose you have a 24VDC nominal battery that is supplying a 4A, 120VAC load with a duty cycle of 4 hours per day. The load is 4A × 4 hours = 16Ah. In order to find the true battery load, divide the load voltage by the nominal battery voltage, and then multiply this times the amp-hours. In this case:

Battery capacity = (load voltage/nominal battery voltage) × Ah

Battery capacity = 120V/24V = 5 × 16Ah = 80Ah

Alternatively, you can determine the total required watt-hours, and then divide by the nominal DC voltage:

Battery capacity = 4A × 120VAC × 4 hours = 1,920 watt-hours/24VDC = 80Ah

Additional capacity can be added to supply the desired number of days of autonomy or battery backup. A typical system is sized for three days of autonomy. To calculate the battery bank size for

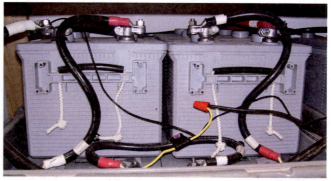

101F30.EPS

Figure 30 Installed batteries.

the desired days of autonomy, multiply the battery capacity by the number of days of required power times the desired DOD. For example, suppose you have a standalone system with a daily load of 80Ah at 24V, three days of autonomy, and a DOD of 80 percent. The size of the battery bank can be determined as follows:

$$\text{Battery bank} = \text{capacity} \times \text{days of autonomy} \times \%\text{DOD}$$

$$\text{Battery bank} = 80\text{Ah} \times 3 \text{ days} \times 80\% = 192\text{Ah}$$

This system could be served by eight batteries rated at 24V each (192Ah/24V = 8).

7.5.0 Selecting a Charge Controller

Properly sized charge controllers help to ensure smooth power and long battery life. The selection of a charge controller depends on the application. PV systems with outputs over 200W often use MPPT charge controllers. They are more expensive but worth the investment in greater efficiency when installed to support larger systems. Smaller systems typically use PWM charge controllers.

Charge controllers must be carefully matched to the size of the battery bank (normally 12V, 24V, or 48V), the AC short-circuit current (I_{SC}) of the solar array, and the maximum system voltage. To size a charge controller, multiply the I_{SC} of the entire array by a factor of 1.25. (Note that this is a minimum value and may vary depending on manufacturer.) The charge controller maximum input voltage must be higher than the maximum system voltage in PWM charge controllers. The design of MPPT charge controllers allows a greater nominal array voltage than the battery voltage. This permits the use of smaller wire between the battery and array.

7.6.0 Adjusting PV Conductors

PV system conductors must be adjusted for many factors. These include the following:

- Temperature
- Continuous duty
- Conduit fill (more than three conductors in a raceway)
- Voltage drop

7.6.1 Temperature Adjustment

Per *NEC Table 310.15(B)(3)(c)*, conduit exposed to sunlight on a rooftop requires an adjustment for the ambient temperature. This adjustment depends on the amount of spacing between the conduit and the roof surface.

Per *NEC Section 690.7(A)*, the maximum PV system voltage is calculated as the sum of the rated open-circuit voltages for all series-connected panels corrected for the lowest expected ambient temperature. The rated open-circuit voltage (V_{OC}) of a PV panel is measured at 25°C (77°F). Temperature correction is required to ensure that the output voltage will not exceed the maximum input voltage of the inverter or the rating of conductors, disconnects, and overcurrent devices. The voltage correction factors are listed in *NEC Table 690.7* and shown in *Table 2*.

The lowest available temperature for a specific area is sometimes known as the winter design temperature, and the month in which it occurs is known as the critical design month. These values

Table 2 Voltage Correction Factors for Crystalline and Multicrystalline Silicon Modules (Data from *NEC Table 690.7*)

Correction Factors for Ambient Temperatures Below 25°C (77°F). (Multiply the rated open circuit voltage by the appropriate correction factor shown below.)		
Ambient Temperature (°C)	**Factor**	**Ambient Temperature (°F)**
24 to 20	1.02	76 to 68
19 to 15	1.04	67 to 59
14 to 10	1.06	58 to 50
9 to 5	1.08	49 to 41
4 to 0	1.10	40 to 32
−1 to −5	1.12	31 to 23
−6 to −10	1.14	22 to 14
−11 to −15	1.16	13 to 5
−16 to −20	1.18	4 to −4
−21 to −25	1.20	−5 to −13
−26 to −30	1.21	−14 to −22
−31 to −35	1.23	−23 to −31
−36 to −40	1.25	−32 to −40

can be found using local weather information or data published by the American Society of Heating, Refrigerating, and Air-Conditioning Engineers (ASHRAE).

To use *Table 2*, multiply the factor for the lowest expected temperature by the rated V_{OC} found on the panel nameplate. For example, suppose a PV panel has a V_{OC} of 20V and four of these panels will be connected in series. This system will be installed in an area where the lowest expected temperature is 0°C (32°F). Per *Table 2*, the correction factor for this temperature is 1.1.0. The system output voltage can be calculated as follows:

$$\text{System } V_{OC} = \text{panel } V_{OC} \times \text{no. of panels} \times \text{correction factor}$$
$$\text{System } V_{OC} = 20V \times 4 \times 1.10 = 88V$$

> **NOTE**
>
> *NEC Table 690.7* is based on crystalline PV panels that have been in use for many years. Newer panels may incorporate alternative semiconductor materials with varying temperature coefficients. Due to these variations, the manufacturer's requirements for temperature adjustment must be used where available per *NEC Section 690.7(A)*.

7.6.2 Continuous Duty

PV loads are considered continuous loads. Per *NEC Section 690.8(B)*, conductors used for continuous duty (operated more than three hours at a time) must be sized at 125 percent of the continuous load. For example, if the load is 20A, the conductor must be rated as follows:

$$\text{Continuous duty load} = \text{actual load} \times 125\%$$
$$\text{Continuous duty load} = 20A \times 125\% = 25A$$

7.6.3 Conduit Fill

Per *NEC Section 310.15(B)(3)(a)*, more than three conductors in a raceway or cable must be adjusted using *NEC Table 310.15(B)(3)(a)*. For example, suppose a 6 AWG USE-2 conductor is to be used with a total of four conductors in a raceway. Referring to *NEC Table 310.15(B)(3)(a)*, the deduction for 4 through 6 cables in a raceway is 80 percent of the values found in *NEC Table 310.15(B)(16)*. A 6 AWG USE-2 conductor is rated at 75A per *NEC Table 310.15(B)(16)*. The adjusted value can be calculated as follows:

$$\text{Adjusted amperage} = \textit{NEC Table 310.15(B)(16)} \text{ amperage} \times 80\%$$
$$\text{Adjusted amperage} = 75A \times 0.80 = 60A$$

7.6.4 Voltage Drop

Excessive circuit lengths can cause a significant decrease in system output voltages. This is particularly noticeable in low-voltage PV systems. In PV systems, voltage drop should be limited to 3 percent or less.

8.0.0 INSTALLATION

The two most common system installations are roof-mounted and ground-mounted arrays. Roof-mounted installations are more common in residential work where the array space on the ground may be limited due to lot size. In addition, roof-mounted arrays are more aesthetically pleasing and this location helps to limit accidental contact with energized components. Ground-mounted installations are often used for commercial systems where access to unauthorized personnel can be limited by restricted areas or fencing. A ground-mounted installation is more time-intensive than a roof installation due to the larger support system. The safe installation of either type of system requires an understanding of the forces exerted on panels and support structures.

8.1.0 Forces Exerted on the Panels/ Support System

Solar arrays are subjected to a variety of forces that can stress the panels and structural components. These forces include expansion and contraction, drag, and wind.

8.1.1 Expansion and Contraction

As a solar panel heats and cools, the metal frame and array mounting structure expand and contract. This change in length is multiplied when panels are racked side by side in a large array. For this reason, any installation in excess of 50 feet in length should include an expansion joint for both the array and the mounting structure or as designed by the professional engineer (PE) of record. The solar panel expansion joint should not straddle the mounting panel expansion joint.

8.1.2 Drag

If an array is installed in an area with heavy snow/ice loading, the solar panels will be subjected to a drag force that pulls the panels at a downward angle. To resist the drag force, the solar panel must have support on the lower edge of the panel frame. Additional mounting clips may be required for systems with excessive drag.

8.1.3 Wind

Wind loads can stress or damage PV panels and supports. In windy conditions, a solar array can exert several thousand pounds of uplift force against the structural supports. Wind loads are particularly critical with angled roof supports as the panels can act as a sail, pulling the array away from the structure.

Careful evaluation of wind loads is essential to ensure a safe installation and is also required to secure certain types of rebates. Wind loads are labeled on support systems and their associated fasteners. Some pre-engineered systems may be supplied with a PE stamp to indicate that they have been evaluated for a specific wind load. All installed systems require PE certification. In addition, systems installed in earthquake-prone areas must be rated to resist seismic loads.

101F31.EPS

Figure 31   Roof layout.

> **NOTE**
>
> Contact the local utility to coordinate the installation of the special metering equipment required with grid-tied and grid-interactive PV systems.

> **WARNING!**
>
> Only qualified individuals may install PV system arrays. Wear all required personal protective equipment when installing PV systems. Apply sunscreen and stay hydrated when working in hot temperatures. Always use fall protection on work surfaces six feet or more above the ground. Do not step on or rest objects on solar panels. Be aware of all electrical hazards.

> **CAUTION**
>
> All roof penetrations, including fastener heads, must be weathersealed to avoid leaks.

8.2.0 Roof-Mounted Installations

An approved fire-rated roofing product must be installed on the roof deck before installation of a PV array. Selection of the roofing product depends on local code requirements, building design, and user preference. In many cases, the fire-rated product consists of asphalt shingles or concrete roof tiles. It must cover the entire roof surface, including areas where PV panels will be installed. In order to minimize shading losses at low sun angles, follow the manufacturer's requirements for panel spacing. For best appearance, center the array on the roof. The roof layout is determined using chalk lines, as shown in *Figure 31*. Roof panels can be laid out in either a portrait (vertical) or landscape (horizontal) orientation.

All mounting holes must be predrilled. To ensure system integrity, all mounts must be attached to structural members such as rafters and trusses. Follow the manufacturer's instructions for tightening fasteners to the proper torque. Secure connections are essential because the wind force and other loads are transferred to the roof structure via the mounting rails. Rails are normally fastened using stainless steel lag bolts since many roofs do not have attic spaces that would allow for access in order to use other types of fasteners. Wind loading can be minimized by mounting arrays no closer than three feet to any edge of the roof surface. Test each panel before installation.

Panels are secured to the roof deck in one of three ways: direct mount, rack mount, and standoff mount:

- *Direct mount* – Direct mount systems mount directly to the roof. These systems provide an unobtrusive appearance and are subjected to the lowest wind loads. However, they prohibit air circulation beneath the panel. This increases the panel temperature, decreasing the output. *Figure 32* shows a direct-mount roof array laid out in a portrait orientation.
- *Rack mounts* – Rack mounts secure the panels using a triangular support. They may be supplied with or without an adjustable tilt angle.

Figure 32 Installed roof array.

Figure 33 Setting footers.

They are commonly used on flat roofs and designed to match the desired tilt angle. These mounts provide the greatest degree of air circulation and also provide easy access to electrical connections. However, they are also subject to the greatest degree of wind loads. When installing rack-mounted arrays, it is essential to follow the manufacturer's instructions regarding inter-row spacing. Improperly spaced arrays will create shadows, seriously reducing or even preventing system output voltages.

- *Standoff mounts*– Standoff mounts are the most common type of mounting system for sloped roofs. They provide three to five inches of space between the panel and the roof. This allows for a reasonable degree of air circulation while minimizing wind loads.

Figure 34 Installing the support system.

NOTE

Some commercial rooftop systems use a ballast (weighted) system rather than traditional roof mounts. Due to weight and wind loading, these systems must be engineered to match the specific structure and location.

8.3.0 Ground-Mounted Installation

When installing a ground-mounted array, the site must first be cleared of any vegetation and leveled. Next, concrete footers are installed to support the array (*Figure 33*). This anchor system requires engineering assistance to ensure sufficient protection against expected loads.

After the footers are installed, the support system is erected according to the manufacturer's instructions (*Figure 34*). *Figure 35* shows a completed support system.

Figure 35 Completed support system installation.

When the support system is complete, the panels are installed. Test each panel before installation. All fasteners must be tightened to the manufacturer's specifications (*Figure 36*). *Figure 37* shows a

Figure 36   Torquing fasteners.

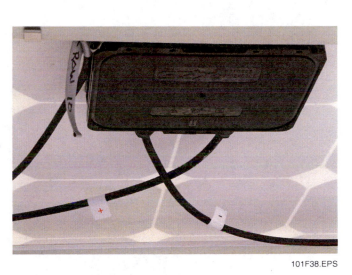

Figure 38   Making electrical connections.

Figure 37   Installed ground-mounted array.

completed ground-mounted array. White marble chips or other light-colored stone is used to reflect light and maximize panel output.

8.4.0 Electrical System Installation

After the array has been installed, the final electrical connections are made. The panels can be joined using traditional conductors and connectors, but are more often installed using quick-connect systems. The panels shown in *Figure 38* use a quick-connect system with sealed termination boxes. The factory wiring is identified for the positive (+) and negative (−) panel output polarities.

Size and install the equipment grounding conductor in accordance with local requirements and the *NEC*®. Attach the equipment grounding conductor to the panel frame using the hole and hardware provided.

Connections are also made to the inverter, batteries and charge controller (if used), panel, and disconnects. Use Type USE-2 copper wire (sunlight resistant and 90°C), for all wiring exposed to weather. The minimum wire size is listed on the panel nameplate. For example, a panel might list the minimum wire size as 12 AWG and the maximum wire size as 8 AWG. (Remember, wire sizes increase as the AWG size gets smaller, then they jump to 1/0, 2/0, and so on.)

Grid-tied systems typically feed into the last two spaces at the bottom in the main service panel. When adding PV power to a panel, do not exceed 120 percent of the bus rating. See *NEC Section 705.12(D)(2)*. Additional capacity may be added when necessary by reducing the size of main breaker or tapping ahead of the bus feed and adding a fused disconnect. Note that all components must be labeled per the *NEC*® and the proper working clearance must be provided for each piece of equipment. *Figure 39* shows the completed electrical installation.

After the system has been installed, it must be tested (commissioned) to ensure that it meets the expected outputs. Check for output voltage, current, and polarity. Program inverters and charge controllers. Review the operation of the system with the building owner and provide all operating manuals for the installed equipment.

Figure 39 Completed electrical installation.

8.5.0 Assessing System Output Power

Assessing the expected output of a grid-tied PV array requires the application of various panel correction factors. Typical correction factors include the following:

- Irradiance (varies between 1–10 percent)
- Array temperature (varies between 1–10 percent)
- Panel mismatch and production tolerances (4–5 percent)
- Dust (7 percent)
- Wiring losses (3 percent)
- Inverter conversion losses (4–5 percent)

Taking into account all of these system losses, the actual array output should be somewhere in the neighborhood of 70–80 percent of the STC value. This means that the measured output of a 1,000W PV system should be somewhere between 700W and 800W on a sunny day.

Assessing the expected output of a grid-interactive system is more complicated due to the addition of the batteries. Grid-interactive systems do not operate at maximum output power unless equipped with an MPPT charge controller. In addition, the batteries must be fully charged and the system must allow full input power to the inverter in order to assess system operation using this method. Many charge controllers and inverters provide system output and efficiency data on a digital display.

9.0.0 MAINTENANCE

While PV systems are generally low on maintenance requirements, there are certain checks and tests that must be performed to ensure proper system performance. The most important factor in the continuing operation of a well-designed system is periodic maintenance. Common quarterly maintenance checks include the following:

> **WARNING!**
>
> Only qualified individuals may perform PV system maintenance tasks. Wear all required personal protective equipment when installing PV systems. Apply sunscreen and stay hydrated when working in hot temperatures. Always use fall protection on work surfaces six feet or more above the ground. Do not step on or rest objects on solar panels. Be aware of all electrical hazards.

- Examine the panels for dirt, dust, sap, or bird droppings and clean using the manufacturer's recommended solution.
- Check panel glass for breakage or cracks.

- Check for loosened, corroded, or burnt electrical connections.
- Observe the condition of all wiring insulation.
- Make sure all grounding clips and wires are intact and secure. Inspect and test the grounding system.
- Check for loosened fasteners. Ensure that all fasteners are tightened to the manufacturer's torque requirements.
- Observe the growth of trees and shrubbery in the vicinity of the array. Have the building owner remove or trim any vegetation that is in danger of shading the array.
- Check the operation of inverters, batteries, and charge controllers. These components are normally replaced one or more times during the life of the system.

10.0.0 TROUBLESHOOTING

PV systems are subject to various environmental stresses, including temperature changes, wind, dirt, dust, lightning, and static electricity. In addition, manufacturing imperfections, improper installation, and equipment age can lead to component or system failures. A variety of tests can be performed to identify and isolate system malfunctions.

10.1.0 Loose or Corroded System Connections

Panel malfunctions can often be traced to a loose or corroded system connection. One way to locate a malfunctioning panel in an array is to shade a small portion of each panel in turn using a piece of cardboard or similar obstruction. This is known as a selective shading test. With the array connected and working, monitor the current. Now, shade a portion of one panel. You should see the current drop. If the current does not drop, then the panel that you are shading is out of the circuit. Look for a fault in the wiring of that panel, or of another panel that is wired in series with it.

10.2.0 Inverter Losses

Inverter losses are often caused by dirty panels. If you observe a drop in the inverter output and it is not due to shading, check the panels for dirt/dust

---

Think About It

Fastener Checks

Some installers use permanent marker to draw a line over the center of the fastener when it is initially installed and after it has been checked for proper torque. That way, they can tell at a glance when fasteners have become loose. Can you find the loosened fastener in this picture?

101SA06.EPS

---

buildup and clean if necessary. Always use the manufacturer's recommended cleaning solution.

10.3.0 Heat Fade

Heat fade can be observed if the system operates inefficiently during periods of high heat. It is usually caused by poor connections or undersized wiring. Heat fade can be confirmed by throwing water on the panels to cool them down while monitoring current fluctuations. Heat fade is more common in grid-interactive and standalone systems due to the addition of the battery system.

10.4.0 Burnt Terminals

Over time, repeated temperature cycling, oxidation, and corrosion may eventually cause enough resistance to burn terminal connections. When repairing burnt terminals, replace all metal parts that have been severely oxidized.

Burnt terminals are usually the result of wiring too many panels in the same circuit. If this is the case, rewire some of the panels on a second circuit.

10.5.0 Bypass Diode Failure

Most PV panels have bypass diodes in the termination boxes to protect cells from overheating during sustained periods of partial shading. Bypass diodes are wired in parallel to prevent the reverse bias voltage caused by partial shading. Diodes may occasionally fail due to lightning or other voltage surges. If this occurs, the diode will normally short out and drop the panel voltage. Replace it per the manufacturer's instructions.

11.0.0 CODES AND STANDARDS

Industry codes and standards are designed to monitor, review, and enforce safety policies and procedures. The PV industry is covered under the following codes and standards:

- Local and national building codes
- International Association of Plumbing and Mechanical Officials (IAPMO): *Uniform Solar Energy Code*

On Site

Microinverters

Enphase microinverters attach directly to individual solar panels using low-voltage DC wiring. This increases system safety and allows for easy expansion.

- Institute of Electrical and Electronics Engineers (IEEE): *IEEE 1547, Standard for Interconnecting Distributed Resources with Electric Power Systems*
- Underwriters Laboratories (UL): *UL Standard 1703, UL Standard for Safety, Flat-Plate Photovoltaic Modules and Panels* and *UL Standard 1741, Standard for Inverters, Converters, Controllers and Interconnection System Equipment for Use with Distributed Energy Resources*
- Occupational Safety and Health Administration (OSHA): *OSHA Standard 1910.302, Electric Utilization Systems*
- National Fire Protection Association (NFPA): *National Electrical Code® (NFPA 70)* and *Standard for Electrical Safety in the Workplace® (NFPA 70E)*

12.0.0 EMERGING TECHNOLOGIES

There are tremendous research efforts in the field of PV systems. Some of the emerging areas of research include replacements for traditional semiconductors, while others involve methods of improving efficiency in solar collection and energy storage.

One new type of system replaces silicon with a light-sensitive dye that absorbs light and produces current. These are known as electrochemical solar cells. When this technology matures, electrochemical cells will be very popular as they are simple and less expensive than traditional solar cells.

Another developing technology uses solar concentrator systems. Solar concentrators use mirrors or lenses to focus light onto specially designed cells. Unlike other PV systems, concentrator systems will not operate under cloudy conditions. They generally follow the sun's path through the sky during the day using single-axis tracking. Single-axis tracking adjusts the vertical tilt of the panels. To adjust to the sun's varying height in the sky through the seasons, dual-axis tracking is sometimes used. Dual-axis tracking adjusts both the horizontal and vertical axes of the panels.

Research in solar energy storage includes the development of household fuel cells to replace traditional battery systems. These differ from traditional fuel cells, which use rare components and are very expensive to manufacture. This process uses sunlight and a special chemical component known as a catalyst to split water into hydrogen and oxygen, which are later recombined in the fuel cell to produce energy. It is clean, efficient, and inexpensive. According to researchers at the Massachusetts Institute of Technology (MIT), it is possible that household fuel cells could begin to replace wire-based energy delivery within the next decade.

SUMMARY

PV power provides a clean, renewable alternative to fossil fuels. It can be harnessed without disrupting the environment and produces no hazardous waste or emissions. It is also quiet, reliable, and requires little maintenance.

Solar cells produce electricity through the use of semiconductors. Many cells are contained in a solar panel, and these panels are electrically and mechanically connected in an array. The current produced by the array is DC current, which is converted to AC current by an inverter. It then travels to the main panelboard, where it is distributed to power lights and appliances in the structure.

Solar energy can be classified into four types of systems: standalone systems, grid-tied systems, grid-interactive systems, and utility-scale solar generating systems. Most residential and commercial customers use grid-tied and grid-interactive systems. These systems provide an advantage in that excess power can be sold back to the utility in the form of credits. In addition, utility and government rebates are available to offset the cost of installing these systems.

In standalone and grid-interactive systems, backup power is provided by a system of one or more batteries. These batteries are matched to the appropriate charge controller to provide grid-quality power.

As this technology continues to develop, it is likely that PV systems will become one of the primary sources of electrical power.

Review Questions

1. Which of the following is an active form of solar power?
   a. designing a building to maximize sunlight exposure in winter
   b. hanging clothes on a line
   c. heating water in an outdoor tank
   d. using a solar-powered calculator

2. When excess energy is sent back into the grid, it is known as _____.
   a. buyback metering
   b. grid metering
   c. net metering
   d. carryover metering

3. The use of solar power is increasing by _____.
   a. 5 percent per year
   b. 10 percent per year
   c. 15 percent per year
   d. 25 percent per year

4. Which of the following is an advantage of a grid-tied system?
   a. Excess energy can be sold back to the utility.
   b. It can provide PV power at night.
   c. It can operate independent of the utility grid.
   d. It provides power when there is a grid outage.

5. A solar boiler is a type of _____.
   a. grid-tied system
   b. grid-interactive system
   c. solar-generating system
   d. standalone system

6. When a motorized system is used to adjust an array to follow the motion of the sun, it is known as _____.
   a. trailing
   b. trolling
   c. tracing
   d. tracking

7. What is the current when the voltage is 120V and the resistance is 10 ohms?
   a. 10 amps
   b. 12 amps
   c. 120 amps
   d. 1,200 amps

8. Which of the following is true regarding series circuits?
   a. If the circuit is open at any point in a series circuit, current will still continue to flow.
   b. A series circuit provides multiple paths for current flow.
   c. A series circuit is a voltage multiplier.
   d. The resistance of the circuit is equal to the sum of the individual resistances.

9. One hp is equal to _____.
   a. 1 watt
   b. 76 watts
   c. 144 watts
   d. 746 watts

10. If an electric heater uses 1,200W for 10 hours, the power consumed is _____.
    a. 1.2kWh
    b. 12kWh
    c. 1,200kWh
    d. 12,000kWh

11. The prefix *kilo* means _____.
    a. 10
    b. 100
    c. 1,000
    d. 1,000,000

12. Wiring two 30V/5A panels in series produces _____.
    a. 30V/5A
    b. 30V/10A
    c. 60V/5A
    d. 60V/10A

13. Wiring solar panels in parallel _____.
    a. increases the voltage but does not affect the amperage
    b. decreases the voltage but does not affect the amperage
    c. increases the amperage but does not affect the voltage
    d. increases both the voltage and the amperage

14. Which of the following locations is likely to have the highest irradiance?
    a. A desert area with significant dust but few cloudy days
    b. A mountaintop with cool temperatures but no dust and few cloudy days
    c. A hot southern location at sea level
    d. A cool but cloudy northern location

15. The *NEC*® requirements for PV installations can be found in _____.
    a. *NEC Article 422*
    b. *NEC Article 450*
    c. *NEC Article 517*
    d. *NEC Article 690*

16. A type of PV panel having a characteristic flaked appearance is probably made up of _____.
    a. monocrystalline cells
    b. megacrystalline cells
    c. polycrystalline cells
    d. amorphous cells

17. The *NEC*® requirements for stationary battery installations can be found in _____.
    a. *NEC Article 480*
    b. *NEC Article 490*
    c. *NEC Article 551*
    d. *NEC Article 625*

18. An advantage in choosing FLA batteries over AGM batteries is that FLA batteries _____.
    a. are less expensive
    b. require no maintenance
    c. do not leak
    d. do not have special venting requirements

19. The best charge controller for use in a cold climate is a _____.
    a. MPPT charge controller
    b. PWM charge controller
    c. shunt charge controller
    d. relay transistor charge controller

20. Each piece of equipment in a PV system requires a disconnect per _____.
    a. *NEC Section 690.1*
    b. *NEC Section 690.7*
    c. *NEC Section 690.15*
    d. *NEC Section 690.41*

21. PV panels are rated up to _____.
    a. 120V
    b. 240V
    c. 480V
    d. 600V

22. PV systems require the use of _____.
    a. Cat I meters
    b. Cat II meters
    c. Cat II or III meters
    d. Cat III or IV meters

23. The total electrical load of a small hunting cabin is most likely to be served by a _____.

    a. 640W system
    b. 1,000W system
    c. 3,600W system
    d. 12,000W system

24. Suppose the sun rises at 6:00 AM and sets at 9:00 PM. What time should you note the position of the sun in order to determine true solar south?

    a. 6:00 AM
    b. 12:00 PM
    c. 1:30 PM
    d. 3:00 PM

25. The default azimuth angle for locations in the southern hemisphere is _____.

    a. 0 degrees
    b. 90 degrees
    c. 180 degrees
    d. 270 degrees

26. During the spring equinox, the angle of declination is _____.

    a. zero
    b. +23.45 degrees
    c. 23.45 degrees
    d. 90 degrees

27. Suppose a customer lives in an area with four peak sun hours per day, uses an average of 36kWh per day during the critical design month, and desires 50 percent replacement of grid power. What is the required system size?

    a. 1,500Wh/day
    b. 2,500Wh/day
    c. 3,500Wh/day
    d. 4,500Wh/day

28. Suppose you have a standalone system with a daily load of 100 Ah, two days of autonomy, and a DOD of 80 percent. What is the size of the battery bank?

    a. 100Ah
    b. 125Ah
    c. 160Ah
    d. 200Ah

29. An array's I_{SC} rating is 100A. The charge controller for this system must be rated at _____.

    a. 100A
    b. 125A
    c. 175A
    d. 200A

30. An expansion joint is required for any array longer than _____.

    a. 30 feet
    b. 40 feet
    c. 50 feet
    d. 60 feet

31. All installed PV systems require a PE stamp indicating the rated _____.

    a. expansion differential
    b. drag
    c. wind load
    d. snow load

32. A flat roof is most likely to use a _____.

    a. direct mount
    b. building-integrated system
    c. rack mount
    d. standoff mount

33. Which of the following is often the result of dirty panels?

    a. Heat fade
    b. Bypass diode failure
    c. Inverter losses
    d. Burnt terminals

34. Which of the following is often the result of wiring too many panels in the same circuit?

    a. Heat fade
    b. Bypass diode failure
    c. Inverter losses
    d. Burnt terminals

35. Which of the following is likely to use a dual-axis tracking system?

    a. a solar concentrator system
    b. a residential off-grid system
    c. a small commercial grid-tied system
    d. a residential grid-interactive system

Trade Terms Quiz

Fill in the blank with the correct trade term that you learned from your study of this module.

1. A unit of energy, usually of electrical energy, equal to the work performed by a single watt for one hour is called a(n) _____.

2. For a fixed PV array, the angle clockwise from true north that the PV array faces is its _____.

3. The measure of radiation density at a specific location is its _____.

4. A junction box used to connect strings of solar panels to create a larger array, and to provide a convenient array disconnect point is a(n) _____.

5. The material that exhibits the properties of both a conductor and an insulator is called a(n) _____.

6. A device that harnesses the energy produced by a chemical reaction between hydrogen and oxygen to produce direct current is a(n) _____.

7. A complete PV power-generating system including panels, inverter, batteries and charge controller (if used), support system, and wiring is called a(n) _____.

8. A rapid switching method used to simulate a waveform and provide smooth power control is a(n) _____.

9. A diode, often used to overcome partial shading, that directs current around a panel rather than through it is a(n) _____.

10. The measure of a location's relative height in reference to sea level is its _____.

11. The thickness of the atmosphere that solar radiation must pass through to reach the Earth is the _____.

12. A low-efficiency type of photovoltaic cell characterized by its ability to be used in flexible forms is considered to be _____.

13. Converting direct current to alternating current requires a(n) _____.

14. When current flows into the grid, it is called _____.

15. A PV cell or panel operating at a negative voltage, typically due to shading, is called a(n) _____.

16. A material to which specific impurities have been added to produce a positive or negative charge is said to be _____.

17. The battery charge controller that provides precise charge/discharge control over a wide range of temperatures is the _____.

18. The device used to regulate the charging and discharging of the battery system to prevent overcharge and excess discharge is the _____.

19. A type of PV cell formed by using thin slices of a single crystal and characterized by its high efficiency is said to be _____.

20. Standardized panel ratings based on a specific operating temperature, solar irradiance, and air mass are referred to as _____.

21. The sun's altitude and azimuth at various times of year for a specific location or latitude band is known as the _____.

22. To determine a location on the Earth in reference to the equator, find its _____.

23. The measure of the amount of charge removed from a battery system is its _____.

24. When a PV system operates inefficiently during periods of high heat, it is usually caused by poor connections or undersized wiring in a condition called _____.

25. A measure of the average height of the ocean's surface between low and high tide and used as a reference for all other elevations on Earth is _____.

26. During outages and after sundown, supplying a PV system with supplemental power that can function independently through a battery bank requires a _____.

27. Large solar farms designed to produce power in quantities large enough to operate a small city are called _____.

28. A grid-interactive system used with other energy sources, such as wind turbines or generators, is called a(n) _____.

29. A PV system built into the structure as a replacement for a building component such as roofing is called _____.

30. A device that maximizes the collection of solar energy by using mirrors or lenses to focus light onto specially designed cells is called a(n) _____.

31. By pouring liquid silicon into blocks and then slicing it into wafers to create nonuniform crystals with a flaked appearance, a type of PV cell called _____ is formed.

32. The equivalent number of hours per day when solar irradiance averages $1,000W/m^2$, also known as peak sun hours, is called _____.

33. The number of days a fully charged battery system can supply power to loads without recharging is its _____.

34. The angle at which the sun is hitting the array is called the _____.

35. A method of measuring power from the grid against PV power put into the grid is called _____.

36. An array-mounting system designed to adjust either the horizontal or the vertical axis of a panel to follow the movement of the sun is a(n) _____.

37. The panel support system, wiring, disconnects, and grounding that are installed to support a PV array is called the _____.

38. The air temperature of an environment is called the _____.

39. The PV system component consisting of numerous electrically and mechanically connected PV cells encased in a protective glass or laminate frame is called a _____.

40. A PV system that operates in parallel with the utility grid and provides supplemental power to the building or residence is called the _____.

41. A temporary decrease in grid output voltage typically caused by peak load demands is a(n) _____.

42. The position of a panel or array in reference to horizontal and often set to match local latitude or in higher-efficiency systems is called the _____.

43. The angle between the equator and the rays of the sun is called _____.

44. The distortion of light through Earth's atmosphere is called _____.

45. An array mounting system designed to adjust both the horizontal and vertical axes of a panel to precisely follow the movement of the sun is controlled by _____.

46. A semiconductor device that converts sunlight into direct current is called a(n) _____.

47. The type of PV cell that replaces silicon with a light-sensitive dye that absorbs light and produces current is a(n) _____.

48. A standalone PV system that typically provides power in remote areas and uses batteries for energy storage as well as battery-based inverter systems is a(n) _____.

49. The volunteer board of renewable energy system professionals that provides standardized testing and certification for PV system installers is the _____.

Trade Terms

Air mass
Altitude
Ambient temperature
Amorphous
Array
Autonomy
Azimuth
Backfeed
Balance of system (BOS)
Brownout
Building-integrated
 photovoltaics (BIPV)
Bypass diode
Charge controller
Combiner box
Concentrator

Declination
Depth of discharge (DOD)
Doped
Dual-axis tracking
Electrochemical
 solar cells
Elevation
Fuel cells
Grid-connected system
Grid-interactive system
Grid-tied system
Heat fade
Hybrid system
Insolation
Inverter
Irradiance

Latitude
Maximum power point
 tracking (MPPT)
Module
Monocrystalline
Net metering
North American Board
 of Certified Energy
 Practitioners (NABCEP)
Off-grid system
Peak sun hours
Photovoltaic (PV) cell
Polycrystalline
Pulse width-modulated
 (PWM)
Reverse bias

Sea level
Semiconductor
Single-axis tracking
Spectral distribution
Standalone system
Standard Test
 Conditions (STC)
Sun path
Thin film
Tilt angle
Utility-scale solar
 generating system
Watt-hours (Wh)

Nicolás Estévez

Chief Executive Officer
Neosolvis Engineering, C.S.P
Mayaguez, Puerto Rico

It's not such a big jump from Superman to solar energy, not to Nicolás Estévez. As a boy, he wanted to save the world wearing a cape, but today he wants to contribute by working on renewable energies, particularly solar. Nicolás believes it's the answer to many of societies' problems.

How did you get started in solar photovoltaics?
As a young person, I wanted to be Superman. Then, because I loved airplanes, I decided I wanted to be a pilot. (The best I did there is becoming a Frequent Flier.) But when I went to the university to study mechanical engineering, I became fascinated by thermodynamics and all things energy. My parents were very environmentally conscious, and though I was born in the US, I grew up in Venezuela and Puerto Rico. I became aware of the various problems society faces. So I did some soul-searching and found that I wanted to work on renewable energies, particularly solar thermal. It seems the best way for me to help society with the skills I've acquired.

Who inspired you to enter such a new field?
My parents. The general upbringing I got from them inspired me to seek out a career in something I feel passionate about and which contributes to the greater good. Solar engineering is what I realized this was. For several years, I was able to study and work in Europe, where renewable energy is a major priority—the number of solar panels on rooftops is amazing.

What do you enjoy most about your job?
Personal satisfaction that I'm doing good for society and the environment. And the variety of challenges, the technical nature, flexible hours, and travel.

Do you think training and education are important in construction?
Yes. It's important to be able to have a strong workforce that is up to date with current technologies and emerging markets.

Would you suggest construction as a career for others?
Yes, we need more people to build today's designs. The opportunities will be there for a very long time.

How do you define craftsmanship?
Just like I define character. Craftsmanship is doing your job to the best of your ability in a productive and safe manner, whether or not anyone will know.

Trade Terms Introduced in This Module

Air mass: The thickness of the atmosphere that solar radiation must pass through to reach the Earth.

Altitude: The angle at which the sun is hitting the array.

Ambient temperature: The air temperature of an environment.

Amorphous: A low-efficiency type of photovoltaic cell characterized by its ability to be used in flexible forms. Also known as thin film.

Array: A complete PV power-generating system including panels, inverter, batteries and charge controller (if used), support system, and wiring.

Autonomy: The number of days a fully charged battery system can supply power to loads without recharging.

Azimuth: For a fixed PV array, the azimuth angle is the angle clockwise from true north that the PV array faces.

Backfeed: When current flows into the grid.

Balance of system (BOS): The panel support system, wiring, disconnects, and grounding system that are installed to support a PV array.

Brownout: A temporary decrease in grid output voltage typically caused by peak load demands.

Building-integrated photovoltaics (BIPV): A PV system built into the structure as a replacement for a building component such as roofing.

Bypass diode: A diode used to direct current around a panel rather than through it. Bypass diodes are typically used to overcome partial shading.

Charge controller: A device used to regulate the charging and discharging of the battery system to prevent overcharge and excess discharge.

Combiner box: A junction box used to connect strings of solar panels to create a larger array, and to provide a convenient array disconnect point.

Concentrator: A device that maximizes the collection of solar energy by using mirrors or lenses to focus light onto specially designed cells.

Declination: The angle between the equator and the rays of the sun.

Depth of Discharge (DOD): A measure of the amount of charge removed from a battery system.

Doped: A material to which specific impurities have been added to produce a positive or negative charge.

Dual-axis tracking: An array mounting system designed to adjust both the horizontal and vertical axes of a panel to precisely follow the movement of the sun.

Electrochemical solar cells: A type of PV cell that replaces silicon with a light-sensitive dye that absorbs light and produces current.

Elevation: A measure of a location's relative height in reference to sea level.

Fuel cell: A device that harnesses the energy produced by a chemical reaction between hydrogen and oxygen to produce direct current.

Grid-connected system: A PV system that operates in parallel with the utility grid and provides supplemental power to the building or residence. Since they are tied to the utility, they only operate when grid power is available. Also known as a grid-tied system.

Grid-interactive system: A PV system that supplies supplemental power and can also function independently through the use of a battery bank that can supply power during outages and after sundown.

Grid-tied system: See *grid-connected system*.

Heat fade: A condition in which a PV system operates inefficiently during periods of high heat. Heat fade is usually caused by poor connections or undersized wiring.

Hybrid system: A grid-interactive system used with other energy sources, such as wind turbines or generators.

Insolation: The equivalent number of hours per day when solar irradiance averages $1,000W/m^2$. Also known as peak sun hours.

Inverter: A device used to convert direct current to alternating current.

Irradiance: A measure of radiation density at a specific location.

Latitude: A method of determining a location on the Earth in reference to the equator.

Maximum power point tracking (MPPT): A battery charge controller that provides precise charge/discharge control over a wide range of temperatures.

Module: A PV system component consisting of numerous electrically and mechanically connected PV cells encased in a protective glass or laminate frame. Also known as a PV panel.

Monocrystalline: A type of PV cell formed using thin slices of a single crystal and characterized by its high efficiency.

Net metering: A method of measuring power used from the grid against PV power put into the grid.

North American Board of Certified Energy Practitioners (NABCEP): A volunteer board of renewable energy system professionals that provides standardized testing and certification for PV system installers.

Off-grid system: A PV system typically used to provide power in remote areas. Off-grid systems use batteries for energy storage as well as battery-based inverter systems. Also known as a standalone system.

Peak sun hours: See insolation.

Photovoltaic (PV) cell: A semiconductor device that converts sunlight into direct current.

Polycrystalline: A type of PV cell formed by pouring liquid silicon into blocks and then slicing it into wafers. This creates non-uniform crystals with a flaked appearance that have a lower efficiency than monocrystalline cells.

Pulse width-modulated (PWM): A control that uses a rapid switching method to simulate a waveform and provide smooth power.

Reverse bias: A PV cell or panel operating at a negative voltage, typically due to shading.

Sea level: A measure of the average height of the ocean's surface between low and high tide. Sea level is used as a reference for all other elevations on Earth.

Semiconductor: A material that exhibits the properties of both a conductor and an insulator.

Single-axis tracking: An array mounting system designed to adjust either the horizontal or the vertical axis of a panel to follow the movement of the sun.

Spectral distribution: The distortion of light through Earth's atmosphere.

Standalone system: See *off-grid system*.

Standard Test Conditions (STC): Standardized panel ratings based on a specific operating temperature, solar irradiance, and air mass.

Sun path: The sun's altitude and azimuth at various times of year for a specific location or latitude band.

Thin film: See *amorphous*.

Tilt angle: The position of a panel or array in reference to horizontal. Often set to match local latitude or in higher-efficiency systems, the tilt angle may be adjusted by season or throughout the day.

Utility-scale solar generating system: Large solar farms designed to produce power in quantities large enough to operate a small city.

Watt-hours (Wh): A unit of energy typically used for metering.

Appendix

SITE SURVEY CHECKLIST

SITE SURVEY CHECKLIST	
☀ **GENERAL INFORMATION**	
Date of Survey:	
Site Name:	
Contact Name:	
Site Street Address:	
City: **State:** **Zip:** **Country:**	
Phone: ( ) **Fax:** ( )	
Email:	

1. ROOF OR OTHER ARRAY MOUNTING SURFACE	
Check boxes or specify in the blank for items below.	
1.01	**Type of Roof Material or Mounting Surface (Specify)**
1.02	**Roof or Mounting Surface Condition**
1.03	**Age**
1.04	**Supporting Structure (e.g. roof trusses)**
	☐　Accessible
	☐　Adequate Strength
1.05	**Roof or Mounting Surface Slope (e.g., 5/12, flat)**
1.06	**Area (Sq. ft.)**
	- Azimuth Direction (degrees E or W of true South)
	- Eave Height (ft.)
	- Ridge Height (ft.)
1.07	**Accessibility to Proposed Array Location**
	☐　Easy
	☐　Moderate
	☐　Unacceptable
1.08	**Potential for Shading Proposed Array**
	☐　None
	☐　Slight
	☐　Unacceptable

2. INVERTER, UTILITY ACCESS, BATTERIES AND ENGINE-GENERATOR (AS APPLICABLE)	
2.01	**Proposed Inverter Location (Specify)**
2.02	**Accessibility to Proposed Inverter Location**
	☐　Easy
	☐　Moderate
	☐　Unacceptable
2.03	**Proposed Battery Location (Specify, if applicable)**
2.04	**Accessibility to Proposed Battery Location**
	☐　Adequate Ventilation
	☐　Adequate Location
	☐　Accessible
2.05	**Proposed Engine-Generator Location (Specify, if applicable)**
	☐　Adequate Ventilation
	☐　Adequate Location
	☐　Accessible

Copyright © Florida Solar Energy Center

101A01A.EPS

☼ RECOMMENDATION

Check the appropriate box below.

☐ Approve site for system installation

☐ Do not approve site for system installation (If site not approved, specify reasons for rejection below:)

☼ SURVEY REVIEWER INFORMATION

Name:

Organization:

Signature: Date:

Please list other committee members reviewing this design:

Name	Organization

SKETCH ROOF AREA AND PROPOSED ARRAY LOCATION (OR ATTACH ON A SEPARATE PAGE)

Available Roof Area (sq. ft.)

101A01B.EPS

Additional Resources

This module presents thorough resources for task training. The following resource material is suggested for further study.

IEEE 1547, Standard for Interconnecting Distributed Resources with Electric Power Systems, Latest Edition. Los Alamitos, CA: Institute of Electrical and Electronics Engineers (IEEE).

National Electrical Code® (NFPA 70), Latest Edition. National Fire Protection Association (NFPA): Quincy, MA.

Occupational Safety and Health Standard 1910.302, Electric Utilization Systems, Latest Edition. Washington, DC: OSHA Department of Labor, U.S. Government Printing Office.

Photovoltaic Systems, Second Edition. James P. Dunlop. Orland Park, IL: American Technical Publishers.

Standard for Electrical Safety in the Workplace® (NFPA 70E), Latest Edition. National Fire Protection Association (NFPA): Quincy, MA.

UL Standard 1703, UL Standard for Safety, Flat-Plate Photovoltaic Modules and Panels, Latest Edition. Camas, WA: Underwriters Laboratories.

UL Standard 1741, Standard for Inverters, Converters, Controllers and Interconnection System Equipment for Use with Distributed Energy Resources, Latest Edition. Camas, WA: Underwriters Laboratories.

Uniform Solar Energy Code, Latest Edition. Ontario, CA: International Association of Plumbing and Mechanical Officials (IAPMO).

Figure Credits

© iStockphoto.com/Pgiam, Module opener

NASA, 101SA01

Mike Powers, 101F02, 101F16, 101F18–101F21, 101F24, 101SA04, 101F33–101F39, 101SA06

Courtesy of DOE/NREL, Credit – Pete Beverly, 101SA02

Antonio Vazquez, 101F03, 101F04, 101F22

Sharp USA, 101F05, 101F12, 101F13 (panel), 101F14 (thin-film solar panel), 101F23, 101F32

eSolar Inc., 101F06

Nellis Air Force Base, 101F07

Topaz Publications, Inc., 101F13 (inset), 101F28, 101F30, 101F31

Photo courtesy of Energy Conversion Devices, Inc. & United Solar Ovonic LLC, 101F14 (flexible thin-film)

Fronius International GmbH, 101F15

Outback Power Systems, 101F17

Schneider Electric, 101SA03, 101F29

The Eppley Laboratory, 101SA05 (mounted pyranometer)

Copyright LI-COR, Inc. and used by permission, 101SA05 (handheld pyranometer)

Solar Pathfinder, 101F25

With permission of Solmetric Corporation, 101F26

Table 2 reprinted with permission from *NFPA 70®, National Electrical Code®*, Copyright © 2010, National Fire Protection Association, Quincy, MA. This reprinted material is not the complete and final position of the NFPA on the referenced subject, which is represented only by the standard in its entirety.

Florida Solar Energy Center® (FSEC®), a research institute of the University of Central Florida, Appendix

NCCER CURRICULA — USER UPDATE

NCCER makes every effort to keep its textbooks up-to-date and free of technical errors. We appreciate your help in this process. If you find an error, a typographical mistake, or an inaccuracy in NCCER's curricula, please fill out this form (or a photocopy), or complete the online form at **www.nccer.org/olf**. Be sure to include the exact module ID number, page number, a detailed description, and your recommended correction. Your input will be brought to the attention of the Authoring Team. Thank you for your assistance.

Instructors – If you have an idea for improving this textbook, or have found that additional materials were necessary to teach this module effectively, please let us know so that we may present your suggestions to the Authoring Team.

NCCER Product Development and Revision

13614 Progress Blvd., Alachua, FL 32615

Email: curriculum@nccer.org
Online: www.nccer.org/olf

❏ Trainee Guide ❏ Lesson Plans ❏ Exam ❏ PowerPoints Other _____

Craft / Level: _____ Copyright Date: _____

Module ID Number / Title: _____

Section Number(s): _____

Description: _____

_____

_____

_____

_____

Recommended Correction: _____

_____

_____

_____

_____

Your Name: _____

Address: _____

_____

Email: _____ Phone: _____

Index

A

AC. See alternating current (AC)
AGM. *See* sealed absorbent glass mat (AGM) batteries
Ah. *See* amp-hours (Ah)
Air mass, 10, 38
Alternating current (AC) disconnects, 14
Alternating current (AC), inverters for converting, 11
Alternating current (AC) loads, powering, 3
Altitude, 20, 38
Ambient temperature, 11, 38
American National Standards Institute (ANSI)
   fall protection standards, 17
   gate ratings, 17
American Society of Heating, Refrigerating, and Air-
     Conditioning Engineers (ASHRAE), 26
Amorphous, 10, 38
Amp-hours (Ah), 12
ANSI. *See* American National Standards Institute (ANSI)
Array, 1, 38
Array orientation, 20-22
Array sizing, 22-23
Autonomy, 24, 38
Azimuth, 20, 23, 38

B

Backfeed, 4, 18, 38
Balance of system (BOS), 9, 38
Balance of system (BOS) components
   combiner box, 14
   electrical system, 14
   footers and support structures, 14, 15
   illustrated, 15
   overview, 13-14
Ballast roof mount systems, 28
Batteries
   charge controllers for, 13
   charging, 12, 13
   corrosion in, 13
   DC, amp-hour requirements, 24
   in electrical system installation, 29
   fuel cells vs., 33
   installation provisions, 11
   life expectancy, 13, 17
   locating, 24
   safety hazards, 17-18
   types of
      flooded lead acid (FLA), 13
      sealed absorbent glass mat (AGM), 13
   wiring in series vs. parallel, 24
Battery backup. *See* autonomy
Battery bank-charge controller matching, 25
Battery bank sizing, 23
Battery capacity, 12, 24
Battery-powered systems, 2-3, 4
Becquerel, Alexander, 1
BIPV. *See* building-integrated photovoltaics (BIPV)
BOS. *See* balance of system (BOS)
Brownout, 2, 38

B (continued)

Building-integrated photovoltaics (BIPV), 11, 38
Burnt terminals, 32
Bypass diode, 32, 38
Bypass diode failure, 32

C

Capacity
   battery, 12, 24
   increasing system, 29
Charge controller
   defined, 38
   electrical system installation, 29
   grid-interactive system, 4
   maximum power point tracking (MPPT), 13, 25, 30, 39
   pulse width-modulated (PWM), 13, 25, 39
   purpose, 3
   selecting a, 24
   UL standards, 32
Codes and standards, PV industry, 32
Combiner box, 14, 38
Commercial systems
   ground-mounted, 26
   roof-mounted, 28
Concentrator, 32, 38
Conductor adjustments
   for conduit fill, 26
   for continuous duty, 26
   for temperature, 25-26
   for voltage drop, 26
Conduit fill conductor adjustments, 26
Continuous duty conductor adjustments, 26
Continuous loads, 26
Continuous wattage, 23
Contraction forces, 26
Converters, UL standards for, 32
Costs, offsetting, 2, 4, 27
Crystalline silicon modules, voltage correction factors, 25
Current (I), calculating, 5-6, 8
Customer interview, 19

D

DC. *See* direct current (DC)
Declination, 20, 21, 38
Deep cycling, 12
Depth of discharge (DOD), 12, 38
Direct current (DC) disconnects, 14
Direct current (DC), inverters for converting, 11
Direct current (DC) loads, powering, 3
Direct-drive systems, 2-3
Direct mount systems, 27
Disconnects, 4, 14, 29
DOD. *See* depth of discharge (DOD)
Doped, 10, 38
Drag force, 26-27
Dual-axis tracking, 33, 38